Hummingbirds

Jewels in Flight

Connie Toops

Voyageur Press

The author gratefully acknowledges the following, who offered assistance in the preparation of this book: Betty Baker, Fred and Laverne Boden, Bill Calder, Mitch Ericson/Perky-Pet Products, David and Linda Ferry, Nancy Newfield, Glenn and Cheryl Olsen, Wally and Marion Paton, Curt and Anna Reemsnyder, Steve and Ruth Russell, Peter Scott, Walter and Sally Spofford, Sheri Williamson, Tom Wood, and Ellie Womack.

Printed in China

First hardcover edition
97   98   99   00   01   9   8   7   6   5

Second hardcover edition
05   06   07   08   09   5   4   3   2   1

Library of Congress Cataloging-in-Publication Data
Toops, Connie M.
Hummingbirds : jewels in flight / Connie Toops
    p. cm.
   Includes bibliographical references and index.
   First Hardcover Edition ISBN 0-89658-161-6
   First Softcover Edition ISBN 0-89658-382-1 (pbk)
   Second Hardcover Edition ISBN 0-89658-576-X
   1. Hummingbirds. 2. Birds, Attracting of. I Title.
  QL696.A558-T66 1992
  598.8'99—dc20   92-19686
                  CIP

Distributed in Canada by Raincoast Books, 9050 Shaughnessy Street, Vancouver, B.C. V6P 6E5

Published by Voyageur Press, Inc.
123 North Second Street, P.O. Box 338, Stillwater, MN 55082 U.S.A.
651-430-2210, fax 651-430-2211
books@voyageurpress.com
www.voyageurpress.com

On the previous page: Male Anna's hummingbird, photo copyright by Hugh P. Smith, Jr.

Contents

Living with Hummers
5

Flying Jewels
9

Birds of a Different Feather
19

Eating Like a Bird
27

Risks and Rewards of Promiscuity
43

A Lichen-Covered Hideaway
51

The Longest Journey
61

Discovering Hummingbird Secrets
71

Meccas of Hummingbirding
79

The Hummer Celebration
95

Attracting Hummingbirds
101

Index
125

References
126

Feisty rufous hummingbirds watch over their territories from inconspicuous perches. Males, which are unmistakable in rust-colored plumage, produce a wing-rattling zizz sound when they fly. (Photo copyright by Charles W. Schwartz)

Living with Hummers

When I think of hummingbirds, I remember one particular summer morning. As a nature photographer, I am often afield before the sun rises over sawtoothed peaks on the California–Oregon border near my home. I was photographing the mountains against the violet glow of the pre-dawn sky. Framing the picture on one side were branches of a mountain mahogany tree. As I peered through the lens, a movement on a limb caught my eye.

At first I thought it was a knot on the gnarled tree. Closer inspection revealed a sleeping hummingbird. For a moment I stood face-to-face with a tiny ball of green fluff that sat absolutely still, shoulders hunched and bill pointing skyward. I backed away, pulled binoculars from my pack, and watched.

Robins began to chirp and male Brewer's blackbirds chased each other from one clump of sagebrush to the next. Tops of distant buttes began to reflect russet hues of the rising sun. The hummer stirred. The little bird stretched its short neck, leaned forward, and ruffled its feathers. Nervously, it glanced from side to side.

The hummingbird was perched facing east, luxuriating in warm rays of sunlight bathing its breast. After a few moments, it opened its left wing and fanned its tail. Then it furled its right wing out and down in a prolonged stretch. The little bird took another quick scan of the sky—left, right, overhead—and suddenly lifted straight up from the branch. It zipped full speed into the rising sun.

I should not have been surprised to find a sleeping hummingbird near my home. During the summer I have seen up to forty Anna's and rufous hummers at once buzzing around our feeders or perched in nearby trees. We guessed our peak population of hummers was about two hundred birds last summer. They begin to arrive about twenty minutes before sunrise. By full light, their aerial antics resemble a game of musical chairs. All of the perches on the feeders are occupied—sometimes by two hummers at once.

Newcomers hover impatiently and twitter complaints to the seated birds. Occasionally they land pompously on the heads, backs, or even bills of perched hummers. There is a constant cacophony of chirps and scolding sounds. Wings slap and tails flair as birds jostle for position. Confrontations often develop into full-fledged duels. Competitors eye each other, then square off beak to beak, rising high above the feeder until one finally drives the other away.

A steady stream of humming visitors arrives throughout the day, but without the urgency of the breakfast crowd. Between meals, hummingbirds perch in junipers and mountain mahoganies bordering the yard. They blink and yawn, now and then readjusting a few out of place feathers with their bills.

They are small, but not shy. If I am doing chores in the yard, especially if I am wearing a touch of red, one may hover so curiously close I can feel the breeze from its wings on my face. They boldly land on me while I change solution in their feeders.

Calliope hummingbirds migrate from Mexico into the mountains of the western United States and southwestern Canada each summer. Weighing less than 0.1 ounce, calliopes are North America's smallest birds. (Photo copyright by Charles W. Schwartz)

From lookout posts on dead branches they track the comings and goings of other hummers. Many are content to watch from the sidelines, their heads swinging back and forth with each new arrival or departure. The feistiest, however, are never still long. These include male rufous hummers, whose wings rattle with a metallic *zizz* when they fly off to chase interlopers.

Male Anna's hummers utter hoarse notes in repetitive series—usually four squeaks and a pause, four squeaks and a pause. Last summer one claimed a hanging basket of fuchsias outside my front door. It stretched its neck to the limit, bill pointing up as it sang. Its throat bobbed and its tail strained with each tiny sound. Like the rufous, dominant Anna's hummers perch high and swoop down on invaders. They whoop a miniature rebel yell as they attack.

At our house, evening feeding frenzies begin fifteen minutes before sunset and continue for half an hour or more. Jockeying for position intensifies until perches are filled and hungry hummers hover nearby like diners waiting for a table in a crowded restaurant. My husband Pat and I sit inside at our own dining table and watch the spectacular aerial ballet, choreographed against the backdrop of a mountain sunset. As lengthening shadows fade into chilly darkness, the satiated birds peel off one by one, silhouetted fleetingly in the reddish glow as they beeline to roosts for the night.

At our elevation, the first blast of cold can arrive in late summer. Somehow the hummers sense the coming chill. Within two days at the end of August last year, half of "our" birds departed for points south. Two weeks later the last few immature rufous disappeared, leaving one lingering Anna's. It departed September 21, the final day of summer. For several days thereafter, I subconsciously listened for the *tick-tick* of tiny voices or the rapid fluttering of wings, but all I heard was the hollow rattle of the wind.

The view from my window seems strangely subdued now, and it will remain so through the winter. But with spring comes the promise of renewed life and the return of my exuberant hummingbirds.

Flying Jewels

In 1526 a Spanish government official, Gonzalo Fernández de Oviedo y Valdés, was asked to write about the natural history of the Caribbean from his diplomatic post on the island of Hispaniola. His book gave Europeans the first printed description of hummingbirds: ". . . no bigger than the end of a man's thumb . . . and of such swiftness in flight you cannot see the movement of their wings," Oviedo wrote. "The colors shine like those of the little birds artists paint to illuminate the margins of holy books."

Nearly five centuries later, hummingbirds continue to fascinate us with colors that glitter like a rainbow and a fast-paced lifestyle that challenges human ability to comprehend. The Cuban bee hummingbird is the tiniest warm-blooded creature in the world. Males, which are minutely smaller than females, measure two inches, including bill and tail, and weigh about 0.07 ounce. Many dragonflies are larger than this. Ruby-throated and black-chinned hummingbirds, common in the United States, are only slightly bigger. Each weighs as much as a penny. It takes 150 average-sized males to equal a pound.

Bee hummingbirds have reached the lower limit of body size. Their metabolisms (speed of life functions) are incredibly fast and energy-consumptive. Were they any smaller, they probably could not eat enough to stay alive.

Rapid lifestyles make hummingbirds unique. For their size, they require the most energy of any warm-blooded animal. Hummingbird body temperatures are higher than other birds'. Their brains and hearts are proportionally larger, and their heart and breathing rates are faster than all other feathered creatures'.

How did hummingbirds become so specialized? Not much is known about hummingbird evolution because no fossil hummingbirds have been discovered. It is unclear whether hummingbirds diverged from songbirds or whether they are more closely related to swifts. Hummingbirds are found only in the New World.

Like swifts, hummingbirds have tapered wings for rapid, maneuverable flight and unique neck muscles so they can turn their heads quickly. Neither has strong feet for hopping or clinging. The order for swifts and hummingbirds is Apodiformes, meaning "no feet." Early naturalists mistakenly believed these birds did not have feet and flew all the time.

Nests of some tropical swifts are suspended under palm leaves, as are those of the hermit group of hummingbirds. Swifts stick their nests together with saliva, rather than the spiderwebs hummingbirds use. Like hummers, swifts lay small clutches of fairly large, elliptical eggs. Young of both groups require a relatively long time in the nest before they fledge. Studies of bone configurations, however, indicate hummingbirds are internally structured more like songbirds. They may have developed from an ancestor that resembled a honeycreeper.

Hummingbirds are the largest family of nonpasserine birds in the New World. By 1990, 338 species of hummingbirds had been discovered.

Hummingbirds live only in the Western Hemisphere. Rufous hummingbirds are among the widest-ranging, inhabiting much of Mexico in the winter but spreading throughout the mountains, meadows, and coasts of western North America in the summer. (Photo copyright by Charles W. Schwartz)

Twenty of these were added since World War II, with six newly described in the 1970s. The total number of species in Trochilidae, the hummingbird family, may still change. Sufficient life history information is lacking for 80 percent of the hummingbirds known.

Their range from Alaska to Tierra del Fuego, Chile, and their diverse forms indicate hummingbirds have been on earth for a relatively long time. Ornithologists believe hummingbirds evolved in tropical forests. Ancestors were probably insect-eaters, shaped and colored much as hermits are. (Hermits are a tropical subfamily of about forty types of dull-colored, forest-dwelling hummers.) As hummingbirds visited flowers in search of insects, they may have sampled nectar as well.

More than 150 types of hummingbirds are found near the equator. Brazil, Venezuela, Columbia, Ecuador, Panama, and Costa Rica are all excellent countries to visit in quest of hummers. Here, birders can rise from lowlands, through mid-elevation forests and fields, to high mountains within a few miles. Each region is colonized by its own community of hummingbird inhabitants.

Since tropical hummingbirds find year-round sources of nectar and insects within small home areas, few of them migrate. Some adventurous species, however, spend summers in distant warm climates where flowers are seasonally abundant, then return to frost-free areas for the winter. The green-backed firecrown summers at the southern tip of Chile. It migrates 1,800 miles to interior South America for the winter. At the other extreme, the rufous hummingbird flies more than two thousand miles from its wintering grounds in Mexico to Alaska and the Pacific Northwest.

Sixteen species of hummingbirds have nested within the continental United States. Fourteen of these are found in the West or Southwest, where mountainous terrain connects southward into Mexico and the tropics. Their ancestors first ventured to suitable summering areas by following the mountain corridor. Through the evolutionary process, these species now take advantage of coastal, desert, and mountain habitats.

The rigorous flight across the Gulf of Mexico or a circuitous route around its shore slows hum-

Hummingbirds probably evolved in the tropics from ancestors that visited flowers in search of insects. Now most hummingbirds, including this female purple-throated mountain gem photographed in Costa Rica, feed primarily on floral nectar. (Photo copyright by Michael and Patricia Fogden)

Southeastern Arizona hosts more hummingbird species than any place in the United States. The largest of these is the sparrow-sized blue-throated hummingbird. (Photo copyright by Sid Rucker)

Hummingbirds are most abundant near the equator. Fiery-throated hummingbirds inhabit mountainous forests of Costa Rica and Panama. They feed on flowers in the forest canopy. (Photo copyright by Michael and Patricia Fogden)

mingbirds from colonizing the eastern United States. Only the ruby-throat is common there. Although individual Bahama woodstars and Cuban emeralds arrive in the Southeast now and then, they do not cross the water in enough numbers at the same time to begin new population centers.

The Ozark, Appalachian, and Adirondack mountains are excellent places to locate summering ruby-throats. Higher elevations of the Rockies host broad-tails and calliopes, while rufous and calliopes are common in the Cascades. Black-chins inhabit lower foothills of both ranges. The California coast is prime habitat for Anna's and Allen's hummers. Southwestern deserts host Anna's and Costa's hummingbirds. Mountainous regions of southern Arizona attract the greatest diversity of hummers in the United States, including blue-throated, magnificent, broad-billed, white-eared, and violet-crowned hummingbirds. Habitats in the United States where hummers do not abound are few: seashores, grasslands, and treeless sage areas of the western Great Plains.

Hummingbirds figure prominently in Native American legends. Hopi, Pima, and Zuni Indians knew hummingbirds nest during the summer monsoon season. Legends of all three tribes refer to hummers as "rain birds." Pimas believed disturbing a nest could provoke disastrous floods. Zunis and Hopis ascribed their words for bee to hummingbird kachinas. Navahos linked hummingbirds with courage and swiftness, qualities of a great leader. For them, hummers held places of honor with wolves and mountain lions as brave spirit creatures.

Although not as prominent as eagles or ravens in masks, totems, and legends, rufous hummingbirds appear in the folklore of Pacific Northwest Indians. Robert Sebastian, a contemporary British Columbian artist of the 'Ksan tribe wrote of "Sahsin," the hummingbird: "Long ago, one of the signals for good luck and good weather to come was the sight of the hummingbird. When hunters prepared for a hunt, they sang Indian songs that would ask the hummingbird to appear, to inspire a successful hunt and healthy game."

Members of the Kwaguithl tribe of coastal British Columbia called the hummingbird *Kwa a-*

Koom-te. To the south, the Squamish of the Puget Sound region associated hummingbirds with bees and wasps, creatures that helped salmonberries mature. Haidas of the Queen Charlotte Islands linked hummers with buzzing mosquitoes. Art Vickers, a contemporary Tsimshian artist, recalled spoken legends of his tribe. "The hummingbird was treated as a very aggressive and powerful individual," he said, "full of energy."

Aztec Indians of Mexico, who watched hummingbirds year-round, called them *huitzil,* meaning "shining one with weaponlike cactus thorn." When Aztec leader Huitzitzil was killed, legend claims his spirit became a hummingbird. Thereafter, Aztecs believed fallen soldiers were transformed to hummingbirds since the tiny birds dueled and practiced warrior skills. The Aztec war god is recognized by a bracelet of hummingbird feathers on his left wrist. Aztec royalty wore cloaks of glittering hummer feathers.

Hummingbird skins are so small that ornaments made from them were reserved for tribal leaders. When they met Puritan colonists, sagamores of northeastern tribes wore ear pendants made of ruby-throated hummingbird skins. It was in New England in the early 1600s, incidentally, that the name "humbird" came into common usage.

Caciques of Taino tribes in the Caribbean also wore hummingbird ear decorations. For them, the bird was a symbol of rebirth. On Caribbean islands, hummingbirds are "doctor birds." Jamaicans recognized the motion sharp-billed hummers use to pierce flowers for nectar as similar to the jab of a doctor using a lancet. Puerto Rican hummingbird nests are thought by natives to cure asthma. Dominicans believe nests can cure earaches, while some Cubans and Mexicans use dried, pulverized hummingbirds in love potions.

Names for hummingbirds reflect their dazzling plumage. Caribbean Indians called them *colibri,* "sun-god birds." The Spanish description *joyas voladores* means "flying jewels." *Lampornis,* the genus name of the blue-throated hummingbird, stands for "lampbird." The genus of the rufous, calliope, and broad-tail, *Selasphorus,* describes a "torch-bearer."

Zuni Indians of the American Southwest called hummers "rain birds." This inlaid silver medallion by contemporary artist J. I. Livingston celebrates the hummingbird's importance in Zuni legend. (Photo by Connie Toops, courtesy of Favell Museum of Western Art and Indian Artifacts)

Naturalist John James Audubon described hummingbird feathers, such as the iridescent plumage of this green violet-ear, as "glittering fragments of the rainbow." (Photo copyright by Michael and Patricia Fogden)

Above: Hopi Indian hummingbird kachinas feature sharp beaks and showy feathers on the head. (Photo by Wendy Shattil and Bob Rozinski, courtesy of Denver Museum of Natural History.) Left: Spanish explorers called hummingbirds joyas voladores, *"flying jewels." The birds' gemlike colors are also represented in common names, such as the Central American fork-tailed emerald. (Photo copyright by Carol Farneti)*

Hummingbirds vary in size from the two-inch bee hummer of Cuba to the eight-inch giant hummingbird. This largest hummer ranges throughout the mountains of western South America from Ecuador to Chile. (Photo copyright by Clayton A. Fogle)

The exceptional beauty of hummingbird plumage is also portrayed in common names, which liken the birds to gems in flight: berylline, ruby-throated, garnet-throated, and amethyst-throated hummingbirds, lazuline sabrewings, and turquoise-throated pufflegs, for example. Among the subgroups of hummingbirds are sapphires, topazes, emeralds, brilliants, mountain-gems, sungems, jewelfronts, and coronets.

In 1775 George Louis Leclerc wrote of hummingbirds in his *L histoire naturelle:* "Of all animated beings, this is the most elegant in form and the most brilliant in colour. The stones and metals polished by art are not comparable to this gem of nature."

Colonists in North America believed a hummingbird landing by chance on a person indicated he or she would soon have wealth and jewels as fine as the hummingbird's plumage. Unfortunately, the lure of the bejeweled bird was too strong for wealthy Europeans who had never seen such creatures in the wild. A few skins and nests were shipped to Europe with reports on flora and fauna of the New World during initial explorations. By the late 1800s, glittering hummingbird feathers were in high demand to decorate jewelry, hats, fans, and dresses. Preserved skins were shipped in barrels from ports in Colombia and Brazil to London and Paris. Thousands were auctioned to fashion designers.

Nineteenth century collectors purchased stuffed hummingbirds to display. Among the most famous were the Duke and Duchess of Rivoli, a French couple for whom the magnificent (formerly Rivoli's) and Anna's hummingbirds were named. European collectors described several new species of hummingbirds from unusual skins they spotted at auctions. A few specimens in European collections do not have living counterparts. They probably become extinct during this age of slaughter.

By 1990, twenty-nine species of hummingbirds were classified as endangered. The list includes the bee hummingbird, the world's smallest bird. Most endangered hummingbirds are tropical residents, such as the marvellous spatuletail that lives in a restricted habitat within one Peruvian valley.

A major reason for the decline of specialized tropical hummingbirds is loss of habitat. Costa Rica, for example, is a progressive Latin American nation in terms of conservation, with diverse habitats preserved in national parks. Yet since 1940, more than half the native forest lands in Costa Rica have been cut. When old-growth rainforests are harvested for timber or cleared for farmable land, more than trees are lost. Flowering vines, orchids, and bromeliads in their branches, and understory plants such as heliconia are wiped out, too. Hummingbirds that depend upon these plants for food and breeding habitat may have nowhere else to go. Clearings created for banana, citrus, or cacao plantations offer entirely different floral components than the forests that preceded them.

A few hummingbird species thrive around plantations, orchards, and suburban gardens. Forest cutting may decrease the numbers of woodland specialists, such as blue-capped hummingbirds or hook-billed hermits, and increase generalists such as fork-tailed emeralds, buff-bellied, berylline, and rufous-tailed hummingbirds.

Conservation practices in Latin America affect hummingbirds seen in the United States. Hummingbirds of the western United States migrate to Mexico. Ruby-throats winter from southern Mexico into Panama. Ruby-throats will survive near plantations, but few hummers find food when forests are replaced by cattle pastures or row crops such as cotton.

Another concern for hummingbird welfare is the wide use of pesticides. Insecticides and fungicides that have long been banned from sale in the United States because of their harmful effects on humans and wildlife are still exported to Latin American nations. Diazinon, which is widely used to kill invertebrates on lawns, golf courses, and agricultural crops in the United States, has been linked to deaths of two dozen avian species including hummingbirds.

The outlook for hummingbirds in the United States is generally good. In the Southwest, violet-crowned and Lucifer hummingbirds have become more abundant in the past half-century. Ruby-throat populations have declined slightly in the Northeast over the last few decades, perhaps due to habitat loss. Like widespread western species, they inhabit such a large area that flower shortages in one region are offset by an abundance of nectar in another. No species of North American hummingbird would be endangered by a single disaster, such as a drought year or loss of specialized habitat. Overall, United States populations remain fairly stable.

Humans demonstrate strong compassion for hummingbirds. We maintain feeders and plant flowers to attract them, and we go to extraordinary lengths to protect them. A few years ago an electric fence manufacturer made red insulators to attach wires to metal posts. Unfortunately, hummingbirds were drawn to the red, inserted their beaks in the insulators, and were electrocuted. When word reached birders, they flooded the manufacturer with calls and letters to change the insulator color. Squads of hummer-lovers trooped along fences of consenting farmers with spray cans, repainting red insulators black or some other nonattractive color.

Since their discovery, the appeal hummingbirds exert on humans has always been strong. Robert Ridgway, who completed a voluminous study of these captivating creatures for the National Museum a century ago, called them, "the most charming element in the wonderfully varied birdlife of the Western Hemisphere, but also, without doubt, the most remarkable group of birds in the entire world."

Birds of a Different Feather

If an animal has feathers, it is instantly recognized as a bird. Among the approximately nine thousand types of birds, hummingbirds are distinguished by adaptations that allow them to hover and fly backwards. A hummingbird's remarkable aerial maneuverability is possible because of feather arrangement, internal wing structure, and tremendously strong flight muscles.

Like most other birds, hummingbirds have ten stiff, thrust-producing feathers, called primaries, attached to the outermost section of the wing. The albatross, a large seabird very efficient at soaring, has forty secondaries on its long forearm. Hummingbirds do not soar. They have short forearms with only six or seven stabilizing secondary feathers. Their tapered, compact wing design makes hummingbirds fast, quick-turning fliers.

A bird stays aloft because air rushes over the convex top of the wing faster than it passes under. Rapidly moving air above the wing exerts less pressure than that below, creating lift.

A downward wing flap generates power to propel a bird forward. Most birds fold their wings and slot their primary feathers to decrease friction on the recovery. This minimizes air resistance while repositioning the wing for the next power stroke.

Unlike birds that bend their wings at joints comparable to the human shoulder, elbow, and wrist, hummingbird wing bones remain rigidly outstretched in flight. Their wings flex only at the shoulder, but they rotate through 180 degrees. In a classic investigation of hummingbird flight, Crawford H. Greenewalt built a small wind tunnel with a feeder at one end, a fan capable of producing various headwinds, and a motion picture camera that recorded hummingbird movements at 1,500 frames per second. His pictures revealed how hummers hover and fly.

To take flight, hummingbirds do not use their legs to spring into the air as most perching birds do. Instead, they beat their wings rapidly back and forth, lifting their bodies into a nearly vertical position. The motion is similar to a swimmer treading water. On the forward stroke, the leading edge of the wing slices through the air. Hummingbirds do not waste the recovery stroke. When the forward beat is finished, hummers rotate their wings at the shoulder joint so the leading edge of the wing faces back.

You can simulate the effect by extending your arm at right angles to your body, keeping your elbow straight, and tracing a figure eight in the air with your hand. Imagine your thumb is the leading edge of the wing. Your palm will be down on the forward stroke and up on the backstroke.

In this way, hummingbirds create enough lift with both wing beats to keep them aloft while hovering. Birds such as kestrels and rough-legged hawks flap their wings to hover, but do so for only short periods. Frigatebirds soar in place, assisted by lift from facing a headwind.

Each hummingbird wing cycle is completed in $\frac{1}{500}$ second. It takes at least three flaps for a

Rather than using their small, relatively weak legs to spring into the air, hummingbirds such as this calliope lift into flight by beating their wings rapidly. (Photo copyright by Charles W. Schwartz)

hummingbird to lift from its perch. If it then desires to fly forward, it tilts its body into a more horizontal plane.

At full speed, the downbeat propels a hummer much as it does other songbirds. But instead of slotting on the recovery stroke, a hummer wing rotates up while fully outspread. You can simulate this motion by extending your arm at right angles with your palm facing down and tracing a vertically elongated oval.

Hummingbirds achieve additional lift on upstrokes by having very large muscles that raise the wing. In other birds, downstroke muscles have the greatest mass. Hummer downstroke and elevator muscles are more nearly the same size. Hummingbird breast muscle mass, when compared to overall body size, is proportionally the largest of all birds. Flight muscles account for a quarter to a third of a hummer's weight. A hummingbird's breastbone is also exceptionally large. It has a wide, elongated keel nearly as long as the chest cavity, providing plenty of space for the flight muscles to attach.

Hummingbirds are singular in their ability to fly backwards and upside down. Slow-motion photography reveals a startled hummer may flip into a backwards somersault and fly for a beat or two with tail up and head down. Then rolling left or right, it will turn upright again. Greenewalt discovered an entire retreat and roll sequence lasted 1/5 second, about as long as it takes to snap your fingers twice. In less frightening situations, a hovering hummer simply shifts into reverse. It tilts the leading edge of the wing to the rear, pulling itself through the air backwards while its body remains in a vertical plane. Like a car in reverse gear, the hummer does not back quickly, but it is highly maneuverable.

Aerodynamics studies reveal a hummingbird's tapered wings, small size, light weight, and whirring method of flight are perfect for controlled aerial maneuvers. Like fast military fighter jets, the cost to hummingbirds for this airborne mastery is tremendously high fuel consumption.

Chickens, which cannot stay in the air for more than a few moments, have relatively small flight muscles made of broad, white fibers. This muscle type gives short bursts of quick energy. In comparison, hummingbirds have dark red flight muscles. These compact tissues are better suited to sustained motion. Within them is an intricate capillary system to bring nutrients and oxygen from the blood and carry away wastes. The muscles are capable of storing fat and lipase, the enzyme that breaks fat into energy-producing components.

A resting hummingbird's heart pulses eight times faster than the average human heart. In flight or during courtship, a hummer's heart may race to twenty beats per second. When it hovers, a hummingbird uses seven times more oxygen than when it is perched quietly. (Oxygen is consumed in the chemical reaction that powers flight muscles.) In the process, excess heat is produced. Hummingbirds are so well insulated inside their feathers that they cannot disperse heat through the skin by sweating, as humans do. Instead, hummingbirds cool themselves by panting. They move hot air, waste carbon dioxide, and water vapor out of the lungs, drawing in cool air and fresh oxygen.

In human terms, a hovering hummer requires ten times the energy expended by a marathon runner. Greenewalt calculated that if a human could exercise at the rate of a hummingbird and could perspire freely, he or she would have to sweat a hundred pounds of perspiration per hour to properly cool the skin.

When hummingbirds zip by, you might suppose these sprites to be among the quickest of all birds. This notion is partly illusion. Compared to a large object, a smaller object traveling at the same speed appears to be moving faster. Hummingbirds are so small they seem to reach very high velocities. In reality, the fastest avian flier is the peregrine falcon, clocked at 175 mph while diving on prey. Swifts can fly at 80 mph while ducks and geese achieve about 50 mph. Chickadees, representative of small songbirds, manage top speeds of 15 mph.

In his wind tunnel experiments, Crawford Greenewalt clocked ruby-throated hummingbirds at 27 mph. Others have reported forward speeds of 40 mph or more (although these may be wind-assisted) and courtship or escape tactics reaching 60 mph.

Only hummers fly upside-down and in reverse. Here a startled male broad-billed hummingbird rolls away from a flower. (Photo copyright by Sid Rucker)

Flight muscles account for nearly a third of this male rufous hummingbird's weight. The strong muscles are well-suited for hovering. (Photo copyright by Charles W. Schwartz)

Slow-motion photography reveals male ruby-throated hummers beat their wings at the rate of seventy strokes per second and female ruby-throats make fifty strokes per second. The difference relates to size, since female ruby-throats normally exceed male weights by about 10 percent. In general, larger birds flap their wings more slowly. The giant hummer, weighing about 0.7 ounce, beats eight to ten times a second. Miniature hummingbirds like the South American amethyst woodstar or the North American calliope weigh less than 0.1 ounce and flap eighty beats per second. In comparison, chickadees beat twenty-seven times per second, pigeons stroke five to eight

times per second, and herons flap only once or twice a second.

Another surprising hummingbird distinction is plumage. Because hummers are so small, the total number of feathers covering them is not great, ranging from 1,450 to 1,650 per individual. Songbirds of chickadee to meadowlark size are cloaked in about 1,500 to 4,500 feathers. Mallards are robed in twelve thousand feathers and swans in more than twenty-five thousand feathers. High feather counts for waterfowl stem from their large size and the need for many down feathers to insulate them while swimming in frigid water. Hummingbird plumage ranks along with water-

Above: A contour feather of a rufous hummingbird (left) is pigmented and reflects a rusty hue. The iridescent feather (right) is colored by interference, the amplification of certain wavelengths of light by minute layers of melanin. (Photo copyright by Connie Toops)

Right: In full sunlight, the gorget of a male Anna's hummingbird glows with the brilliance of hundreds of rubies. (Photo copyright by Wendy Shattil/Bob Rozinski)

By tilting the leading edges of its wings to the rear, a broad-billed hummingbird backs away from a flower. (Photo copyright by Carl E. Stevens)

fowl in its insulating ability. Proportional to size, hummingbirds have a huge number of feathers, which helps maintain high internal temperatures.

Hummingbird feathers are differentiated into smooth feathers that aerodynamically streamline the body contour and stiff flight supports on the wing and tail. Hummers are distinguished by decorative feathers that sparkle like sunlight dancing through a prism.

Colors of trees, grass, and most other objects surrounding us daily originate in pigments. When light strikes these objects, chemical molecules in the pigments absorb or reject wavelengths of light. Tree leaves and grass appear green because pigments on their surfaces absorb all colors of light except green. Green wavelengths bounce from leaf surfaces to receptor cells in our eyes.

Compared to a rainbow or the shimmering surface of a soap bubble, colors of most pigmented objects seem dull. Rainbows and soap bubbles are examples of structural color rather than light reflected randomly from one surface. They derive from white light split by water drops or soap film into its various components.

In the case of a soap bubble, or of a thin sheen of oil floating on water, colors result from an even more complex process, interference. Here, light waves reflect from two surfaces at once. Color wavelengths from each surface synchronize and reinforce each other's brilliance. The glittering feathers of a hummingbird are also colored by interference.

Bird contour feathers are subdivided into hairlike barbules. In most species, barbules are arranged much as the hook and pile of Velcro, holding the bird's feather covering together. Glitzy hummingbird feathers, which are iridescent only on the outer third, lack the hook and pile zipping mechanism. Instead, each barbule contains stacks of microscopic elliptical plates. These color disks measure about $1/10,000$ by $4/10,000$ inch. Each contains a layer of air sandwiched between thin films of dark-colored melanin. The rainbow of hummingbird colors—from the rosy red of an Anna's or the carmine of a broad-tail's gorget to the stoplight green of a magnificent and the rich violet of a black-chin—is created by minute differences in the thickness of melanin films and the spacing of air layers between them. When light strikes a feather, most frequencies are eliminated because wave peaks cancel troughs of other waves.

Certain frequencies, however, synchronize and strengthen as they bounce off the layers of melanin. They project back from the platelets in a beaconlike beam that can only be seen from a narrow, head-on angle. From that viewpoint, the hummer's gorget seems to glow with the brilliance of the sun on hundreds of rubies or emeralds or sapphires. When a male hummingbird in full sunlight turns toward you, it is almost as though you are momentarily bathed in the beam of a laser. The effect is intensified because surrounding unlit feathers appear black. Shading will change slightly—from red to orange to gold, or green to turquoise—depending on the viewing angle.

Adult males wear the most intense colors. Females, who find advantage in blending into the forest surroundings, are cloaked in drabber hues. The back feathers of both males and females often contain green iridescence, but it is not as brilliant as gorget and crown colors. Platelets on back feathers are concave instead of flat, scattering light across a wider area but with less intensity. Brown and russet body colors, such as those on forest-dwelling hermits or rufous hummingbirds, come from pigmented feathers.

Some hummingbird feathers are modified to produce sounds. Handsome male Jamaican streamertails are glossy green on the back and breast with black cap, wing linings, and tail. They are similar in size to Anna's hummingbirds except that adult males have two trailing feathers that extend six inches beyond the rest of the tail. These specialized feathers curve near the ends and are scalloped along the inner edges. As the birds fly, the feathers produce a soft hum. The tail is displayed prominently in courtship.

Outer tail feathers of the male Anna's hummingbird are slightly curved and a bit stiffer than the inner ones. They can readily be heard during courtship, when males dive to impress waiting females. They pull out of the plunge just above ground level, and in the process generate a popping noise with the specialized feathers.

Wing, rather than tail, feathers are modified to produce the metallic buzz heard when an adult male broad-tailed hummingbird flies. The outermost primary feathers on both wings come to sharp points that vibrate in flight. Adult male rufous and Allen's hummingbirds have similar, but softer, wing-buzzing noises.

Considering the relatively small number of feathers hummingbirds possess, their differentiation is remarkable. In addition to normal functions of flight and insulation, hummers have dazzling display colors and identifiable audible signatures produced by feathers. Hummingbird feathers have evolved to meet dual functions of attraction and deception, responding to diverse roles of enticing mates and avoiding predators.

Although hummingbirds obtain much of their nutrition by sipping nectar from flowers, they will also lap sugary liquid flowing from holes sapsuckers drill in trees. (Photo copyright by Robert and Jean Pollock)

Eating Like a Bird

The phrase "You eat like a bird" is meant as a compliment, implying you have a tiny appetite. Few realize the adage is wrong. Most birds eat great quantities of food, and hummingbirds are no exception. The smaller the bird, the more rapid its metabolism. Hummingbird metabolic rates are more than double those of wrens and twenty-five times as fast as chickens'. To equal a hummingbird's metabolism, a human would need to digest and process twice his or her weight in food every day.

Hummingbirds turn nectar into food energy. Using a slim glass tube, scientists have drawn samples of nectar from flowers hummingbirds visit frequently. They analyze samples with a refractrometer to determine each nectar's sugar content. In a study of more than two hundred flowers favored by hummingbirds in the United States and the tropics, the average sugar content was 25 percent.

Hummingbirds have taste receptors on their tongues and salivary glands. They return to feeders and flowers that offer sweet rewards but reject nectar that is less than 12 percent sugar (the sweetness of Coca-Cola). They also seem repelled by the taste of a few plants, including oleander, which is poisonous to humans.

Hummingbirds lap nectar with their tongues, at the rate of about thirteen licks per second. The tip of a hummingbird's tongue is forked. In many species the outer edges of these tips are fringed to sweep up nectar and tiny insects from inside the flower tube. Nectar is not sucked, in the manner of drinking with a soda straw, but wicked up a shallow, W-shaped groove in the tongue.

At rest, the tongue fits within a hummingbird's beak. Its base is attached to thin tissue behind the throat that winds around the back of the head. With the bill open slightly, this tissue extends to flick the tongue forward. The tissue contracts and the nectar-filled tongue returns. As the bill closes, the tongue is forced out again, and with nowhere else to go, the nectar is squeezed back into the oral cavity.

Because the crop, where food is stored for digestion, is tiny, hummingbirds consume numerous small meals rather than a few big ones. It requires about five minutes from the time a hummer fills its crop with nectar until the contents pass into the stomach. Hummers eat, rest, then eat again. The routine is repeated at least sixty times throughout the day. Hummingbirds spend 10 to 15 percent of their time feeding and 75 to 80 percent perched while digesting.

It takes about nine hours for human food to pass through the stomach and small intestine plus a day or so before wastes are eliminated. Typical of fast-paced hummingbird life, their digestive cycle requires less than an hour from intake of food to conversion into carbohydrate energy.

Although nectar gives hummingbirds a source of quick energy, they need supplemental proteins, vitamins, and minerals. Hummingbirds look closely at the stems and leaves of flowers as they feed,

Overleaf: Flowers attractive to hummingbirds often are red, tubular, and spaced far enough apart to accommodate buzzing wings. Here an Allen's hummingbird samples the nectar of red delphinium. (Photo copyright by Hugh P. Smith, Jr.)

Hummingbirds cannot see potential predators when they dip their heads deep into a flower tube. Trumpet creeper flowers offer more nectar than most other blossoms, so hummers such as this male black-chinned will risk blocked vision in return for an abundant reward. (Photo copyright by Connie Toops)

gleaning spiders and aphids. They scan tree bark, branches, and cones for other insects. In suburban areas, they hover along walls and under eaves, raiding spider's webs. Hummingbirds also sit on bare branches and watch for swarms of gnats or other flying insects. They dart into these swarms, rapidly snapping up and swallowing prey. Hummingbirds eat insects so small the prey is difficult for humans to see. Yet by their erratic flights and jerky head motions, we know hummingbirds are dining when they engage in this fly-catching behavior.

Entomologists have sampled the atmosphere from ground level to a hundred feet high in the midwestern United States in summer. The air above one square mile may contain as many as 45,000 airborne balloon spiders, 324,000 tiny thrips, 813,000 aphids, and seven million fungus gnats. Peak evening hours may include as many as 980 million midges per square mile. Tremendous numbers of these basic links in the food chain of hummingbirds and songbirds are killed by indis-

criminate spraying for mosquitoes and other pests.

The numbers and types of insects eaten by hummingbirds vary from species to species. Lucifer hummingbirds, which have slightly down-curved beaks, often extract insects from flower tubes. Starthroats and blue-throated hummingbirds hawk fairly large airborne insects. Stomach samples from blue-throats have contained remains of flies, wasps, beetles, and spiders. Central American hermits pluck insects from the undersides of leaves. Hummingbirds eat more insects during rainy periods and when flower nectar is scarce.

Walter Scheithauer contributed greatly to knowledge of hummingbird diets with his study of a captive white-ear. First, the bird was offered dextrose and fructose in water with a drop of vitamins added. During sixteen hours of observation the white-ear consumed 80 percent of its weight in sugar and more than eight times its weight in water. It visited the feeder 172 times, drinking an average of five seconds each and resting no longer than twelve minutes between drinks. The bird restlessly searched for insects in its glass-fronted cage but found none. By late afternoon, its eyes were glazed and it was listless.

For the next few days the white-ear received the sweet vitamin solution plus all the water and fruit flies it wanted. It caught an average of 677 fruit flies per day. The hummer consumed 73 percent of its weight in sugar and 27 percent of its weight in insects daily. It remained active and healthy.

Scheithauer ended the experiment by offering the white-ear the sweet solution fortified with blended bananas, powdered milk, egg yolk, mealworms, and blood serum. He continued to release unlimited numbers of fruit flies. The white-ear consumed eight and a half times its weight that day—including the normal amount of water, the normal 70 percent of its weight in sugar, plus an extra 0.14 ounces of proteins and carbohydrates imbibed with it. Scheithauer observed at the end of the day the hummer looked "like a man who has celebrated too well and wants only to rest, sleep, and digest his heavy meal."

The usual daily intake of sugar and insects equals about half a hummingbird's weight. In comparison, humans consume little more than 1 percent of their total weight daily. Research on wild Anna's hummingbirds indicates they derive slightly more than 0.07 ounce of sugar each day from visits to nectar-producing flowers. A Costa's hummingbird in southern California was documented for six and one-half hours in which it made forty-two feeding flights and visited 1,311 individual flowers.

To determine how many blossoms one hummer must visit to satisfy its appetite, scientists have placed net bags over flowers for a day, allowing nectar to collect. Blooms of hummingbird flowers produce from 0.000035 to 0.0001 ounce of sugar daily. Birds the size of ruby-throats or Anna's require about 0.07 ounce of sugar per day. If each blossom achieves minimal sugar production, hummers would probe two thousand flowers per day to satiate their hunger.

But other hummingbirds or insects may already have harvested some of this nectar. The blossom may contain only half its nectar, or per-

haps none at all. It would not be unusual for a hummingbird to explore three thousand or more blooms each day to satisfy its appetite.

Since hummingbirds roam widely in their search for food, they may probe species such as bluebonnets that offer no nectar. The same bird seldom repeats a visit to a nectarless flower. Instead, hummingbirds learn the color, shape, and taste of rewarding flowers. They have excellent memories, especially of colors and locations. Hummers will return from day to day and year to year to rich stands of wildflowers. Rufous hummingbirds were seen around Mount St. Helens in Washington state after it erupted in 1980, presumably trying to locate flowers they remembered from previous summers.

A generic description of a blossom attractive to a hummingbird would include red color, tubular shape, and ample sugar content. Flowers must be open during the day and hang far enough apart that the bird's buzzing wings will not bump the plant as it feeds. Hummer favorites are generally perennial herbs and shrubs, found year after year in the same place and blooming through a fairly long period.

Hummingbirds are efficient pollinators when flower tube lengths and reproductive parts correspond to bill size and shape. Through time certain types of birds and blossoms have evolved together in mutually beneficial relationships. Perhaps as ancestral hummingbirds gathered bugs from flowers, they tasted sweet nectar and gradually depended upon it for food. In the process, their bills or heads may have been dusted with pollen at one plant and transferred it to another.

It is beneficial to the new generation of seeds if the pollen to fertilize them comes from two separate plants. From the plant's perspective, warm-blooded birds are more active, reliable pollinators on cold or rainy days than insects are. If a flower offers a small amount of nectar, a hummingbird will visit briefly. The hungry bird will likely continue to several more blossoms before satisfying its

Above left: Hummingbirds conserve energy during the night by entering torpor, a short-term form of hibernation. (Photo copyright by Willard E. Dilley)

Above: Most insects land on flowers to probe for nectar. Sphinx moths are an exception. They may be mistaken for hummingbirds as they hover to feed. (Photo copyright by Steve and Dave Maslowski)

Left: A few kinds of flowers, including some types of Indian paintbrush, host tiny mites. These smaller-than-a-pinhead-sized creatures hitch rides between plants by crawling up the bills of visiting hummingbirds. (Photo copyright by Wendy Shattil and Bob Rozinski)

appetite, transferring pollen as it goes.

If production of more or sweeter nectar lures better pollinators, the expenditure is worthwhile. Otherwise, energy necessary for nectar production could be invested in stronger plant growth or more seeds. A survey of 156 flowers pollinated by bees showed their average nectar content was 42 percent sugar, sweeter than that needed to attract hummers. Hummingbird flowers, however, must produce greater volumes of nectar because the birds consume more nectar per visit than bees do.

Bees see blue, purple, and ultraviolet light but do not discern red well. In North America the mint and penstemon flower families have numerous blue species pollinated by bees. Although hummers will readily visit rewarding blue flowers, many mints and penstemons serviced by hummingbirds are red. To migrant hummers, red flowers may project the same message a cafe sign flashes to weary motorists on a lonely highway. Red flowers are also relatively free of competition from bees. Botanists have identified about 150 plants in the United States modified through evolution to be pollinated by hummingbirds rather than insects.

Bees, flies, and butterflies are attracted to flowers with odors. Fragrance, which is a separate substance from odorless nectar, is produced at an energy cost to the plant. Hummingbirds do not rely as much on smell, nor do hummingbird flowers usually have strong odors.

Many insects need a lip on which to alight while they probe for nectar. Most hummer-flowers do not offer perching platforms since hummingbirds hover to feed. Downward-pointing openings and long corolla tubes also exclude insects. Hummingbird moths are an exception. These large sphinx moths hover while extracting nectar from flowers with their long tongues. They are sometimes mistaken for hummingbirds in dim light.

Hummingbird flowers are usually pollinated as the bird's head touches reproductive organs dangling above or extending beyond the open tube. Shorter blossoms, such as manzanita and twinberry, are fertilized as the tip of the bill brings pollen into the corolla. In tests on captive hummingbirds in an aviary, pollen encountered on one

flower can be spread over about ten succeeding flowers. Hummingbirds do not like pollen on their beaks. Immediately after feeding they scrape their bills on tree branches, trying to wipe away this sticky substance, which has little or no nutritive value.

Several plants have evolved thick flower walls so they will not be pierced by thieving hummers with rapier beaks. Others have grooves to steer hummers toward nectar but away from fragile ovaries. Penstemons, for example, are shaped so the delicate seed-bearing organs are separated from the nectar reservoir.

Hummingbirds are reluctant to extend their heads so far into flower tubes that they cannot see. Yet hummingbirds will submerge themselves in the cavernous orange flowers of trumpet creeper. The reason? Trumpet creeper produces ten times the nectar of average hummer flowers. The sweet prize of so much sugar in a single location makes it worthwhile for hummers to dip into a flower that blocks their vision of potential predators. If trumpet creeper flowers offered less reward, they would probably be bypassed.

Hovering is energy-consumptive. The only way a hummer's energy budget can afford hovering flight is for the bird to feed on sugars that are quickly converted to energy. Unlike birds of prey who transform three-quarters of the meat they eat, or seed-eating birds who recoup only half their meals, nearly all the sugar consumed by hummingbirds is turned into energy.

About a hundred types of hummingbird flowers host mites. Easily overlooked since half a dozen could fit on the head of a pin, the first hummingbird mites were not discovered until 1969. In tropical lowlands, where host plants bloom year-round, mites stay on one kind of plant. Tropical mountain habitats experience more seasonal change. Here, mites may live on two or more hosts that bloom at different times.

The only hummingbird mites so far identified in the United States summer on Indian paintbrush in coastal California. They winter in west-central Mexico. How do mites move between flowers? They hitch rides in the nasal cavities of hummingbirds. Numerous tropical hummers, plus Anna's,

Left: Some tropical hummingbirds have distinctly curved bills that coincide with the shapes of specialized flowers. Here a white-tipped sicklebill feeds on Heliconia in a Costa Rican rainforest. (Photo copyright by Michael and Patricia Fogden)

Above: Ocotillo flowers in southwestern Texas are often "robbed" of their nectar by carpenter bees that pierce the floral tubes. Hummingbirds do not linger around nectarless blossoms. In southern California and Arizona, where carpenter bees seldom steal from ocotillo, black-chinned, Costa's, and Anna's hummingbirds readily lap nectar from flowers of this desert shrub. (Photo copyright by Sid Rucker)

Allen's, and rufous in the United States, are known to carry these harmless creatures.

Female mites lay pinpoint-sized eggs on host plants. Within a week, eggs hatch and develop from larva into adults. Like hummingbirds, they feed on nectar. Mites abandon a flower as the bloom fades, milling about on the stem until another opens.

To reach a separate plant, a mite senses the wing whooshes of a feeding hummingbird. In the brief seconds of its visit, the mite races up the hummer's beak into its nostril, which will already have other mite passengers. Each species waits for the faint aroma of its host plant to waft into the nostril as the hummingbird makes its rounds. When the odor is right, the mite has about five seconds to scurry down the beak to its new home. On that specific host, the mite will find mates. Amazingly, mites have adapted their lifestyles to fit pollination rituals of their hosts. They are not known to injure plants or hummers.

Hummingbirds exploit several animal-produced sources of nectar. In Colombia and Brazil, bark-dwelling insects secrete a sugary substance. Tropical hummingbirds visit trees where these insects live to feast on the sweet residue.

In the eastern United States, yellow-bellied sapsuckers drill spiralling sets of holes around the trunks and large branches of hardwood trees. In the West, red-breasted, red-naped, and Williamson's sapsuckers employ similar strategies. The shallow wounds drip with sap that has been disrupted from its normal flow through the tree's cambium layer. The sticky fluid, varying from 15 to 25 percent sugar, attracts gnats and other insects. Sapsuckers return periodically to harvest both insects and syrup.

A Michigan study showed ruby-throated hummingbirds fed at sapsucker works more frequently than the makers of the holes or any of the nineteen other species of birds that sometimes visit these wells. In northern New England and Canada, ruby-throats arrive several weeks before wildflowers began to bloom. They survive by following sapsuckers and feeding at their drillings. The northern limits of the ruby-throat's range correspond to northernmost sapsucker populations.

A birder in the Adirondack Mountains reported: "An early morning walk brought us to a grove of paper birch trees where . . . a number of the birch trees bore the telltale signs of repeated annual drillings by sapsuckers. As we watched, a male downy woodpecker came to feed. As the downy feasted on the sap, a female hummingbird investigated the woodpecker's red head markings. Then when the downy left, it was the hummer's turn at the sap."

Hummers can perch near sapsucker wells, saving energy in comparison to hovering at widespread flowers. Individual rufous hummingbirds will defend sapsucker trees as their personal territories. In Colorado, rufous and broad-tailed hummingbirds have been observed dueling over sapsucker drillings in aspen and willow groves.

When two or more species of hummingbirds compete for limited nectar, they may develop strategies to insure each receives its share. Broad-tailed hummingbirds nest in the Rocky Mountains of Idaho, Wyoming, Colorado, and northern New Mexico soon after Nelson's larkspur begins to blossom. This is an early-blooming species with three to ten purple, tubular flowers per stalk. Broad-tails residing near wetter areas also use red Indian paintbrush and clumps of subalpine larkspur. As the larkspurs taper off, trumpet-shaped scarlet gilia blooms. The flowering of these plants coincides with the period in which broad-tails breed and raise chicks.

Male broad-tails abandon their territories in mid-summer and are often replaced by southbound migrant rufous hummingbirds. Rufous hummers are quite aggressive. Even in migration, they select small, robust stands of wildflowers, like scarlet gilia, that can sustain their energy needs. They may remain in one spot for several days, defending the flowers against all comers.

Tiny calliope hummingbirds also migrate south at this time but they shy away from dog-

fights with rufous and broad-tails. Instead of depending on one rich stand of flowers, calliopes zip from plant to plant across a wide area. They travel a circuit of two to three hundred blossoms. From a flower's perspective, calliopes are excellent agents of cross pollination. Calliopes keep a low profile, foraging on bottom blossoms rather than conspicuous flowers where they might be spotted by dominant rufous or broad-tailed hummers. Calliopes also visit higher elevations to find flowers not in use by other species.

Bill Calder, a professor of ecology and evolutionary biology at the University of Arizona, has an electronic balance connected to a computer to automatically weigh hummers visiting a feeder at his summer banding station in the central Colorado Rockies. Male territory owners come often but take only a few sips each time. Nesting females, who are preoccupied with raising young, slip into other birds' territories rather than defending their own. They visit the feeder less frequently but drink more at each fill-up. Males sneaking in from outlying territories consume an intermediate volume—enough to pay for the energy spent reaching the feeder but not drinking so long or so much to prevent a fast getaway if challenged.

While studying trans-Gulf migrant songbirds on a Mississippi barrier island, ornithologist Paul Kerlinger noted a difference in feeding behaviors of male and female ruby-throated hummingbirds. Males arrived in late March, two weeks earlier than females, and established territories around large thistles blooming near freshwater marshes. They defended these against other hummers as well as larger, nectar-seeking orioles.

Females segregated to scrub and pine woods habitats, where they fed at greenbriar blossoms. Kerlinger watched a few probe prickly pear cactus flowers. Only after male hummers migrated farther north did females venture into the marshes where the thistles bloomed.

Segregation of males and females into different habitats has been documented for numerous other species. It reaches an unusual extreme in saw-billed hermits. Males and females have differently shaped beaks to take advantage of separate nectar sources.

Hummingbirds use long tongues to lick nectar from deep inside flower tubes. (Photo copyright by Charles W. Schwartz)

In the Central American cloud forest, stripe-tailed hummingbirds use short, sharp bills to pierce thin flower tubes. (Photo copyright by Michael and Patricia Fogden)

Hummingbirds have adapted to life in meadows and woodlands, deserts and swamps, tropical lowlands and mountains. In each area, diverse beak sizes and shapes influence floral choices. Bird bills are bony, toothless extensions of the skull and are covered with protective material similar to human fingernails.

The Colombian purple-backed thornbill has the smallest bill of all hummers, a tack-length, 5/16-inch beak. It also has stronger feet than most hummingbirds. It perches to pierce flower tubes, then sips nectar. Short-billed hummingbirds visit small flowers. Hummers with longer, straight bills (typical of most United States species) are generalists. They successfully extract nectar from a variety of floral tube lengths. In the tropics several hummingbirds have very long or distinctly curved beaks. They obtain nectar from specific varieties of flowers that coincide with the shapes of their bills.

The swordbill, whose spiked, five-inch beak is the longest of all hummers, uses it to reach nectar deep in bell-shaped Datura flowers in the

Andes and in long corollas of South American passion flowers. Compatibility of bill and flower shape is more important in the tropics than the color of the flower.

Swordbills are "trapliners" rather than defenders of territories. Since they have little competition from other species of hummingbirds for Datura or passionflower nectar, they commute many miles per day from plant to plant. Trapliners are strong distance fliers and hover efficiently while feeding. Hummingbirds that defend territories are more agile and can accelerate to greater speeds for short aerial chases, but they do not hover as efficiently or range as widely as trapliners.

The spreading blooms of scarlet passionflower, found in equatorial latitudes, have narrow collars in the center to protect the nectar. Few birds other than the long-tailed hermit can reach it. This large bird, found from southern Mexico to the Amazon Basin, possesses a down-curved beak. The white-tipped sicklebill is another specialist. Its beak hooks down at a ninety-degree angle. It

Hummingbirds have excellent memories and return to abundant patches of flowers—or feeders—year after year. (Photo copyright by Clayton A. Fogle)

clings creeper-style to stout heliconia blossoms and probes deep into the curved flowers for a high-energy snack.

Mountain avocetbills of Colombia and Ecuador and fiery-tailed awlbills have beaks that curve up at the tips. The tooth-billed hummer of northwestern South America has a narrow, primitive-looking beak with rough serrations along each edge. The "teeth" may be helpful in catching insects.

In the tropics, groups of specialized hummingbirds coexist in various levels of the understory and forest canopy. Studies near Monteverde, Costa Rica, revealed the green hermit, with its two-inch, decurved bill, pollinates three-quarters of the long-tubed flowers near ground level. A competitor, the stripe-tailed hummingbird, pierces papery tubes of Acanthus flowers to steal nectar. This medium-sized hummer reaches in through the flower wall with a short, sharp beak. Theft does not work with wild plantain flowers, whose waxy leaf cups are hard to puncture. Inside, the

plantain nectar tube is surrounded by a moat of stagnant rainwater. Only the green hermit and a few other long-billed hummers can dip into that store of nectar.

In the same region, more than 90 percent of the short-tubed flowers are visited by the purple-throated mountain-gem, whose beak is less than an inch long. High in the canopy, where small blossoms of a plant related to the blueberry dangle, the fiery-throated hummingbird feeds with less competition from aggressive mountain-gems or hermits below.

In fields around Monteverde, most of the fourteen common hummingbird species have bills of generalists. Behavior plays a large role in determining who feeds where. The shy, nervous fork-tailed emerald starts its rounds at first light, visiting a shrub in the coffee family that has abundant orange flowers.

As the sun rises, nectar production increases. A pugnacious rufous-tailed hummingbird arrives. This russet and green bird is an inch larger than the

Grains of pollen stick to a hummingbird's head and bill as it feeds. (Photo copyright by Charles W. Schwartz)

emerald, which it stubbornly drives away. The rufous-tail steadily consumes nectar over the next hour or so as production peaks. When the rufous-tailed hummer is satiated, a feisty steely-vented hummingbird takes its place.

In the meantime, a tiny magenta-throated woodstar, which looks and sounds like a large bee, uses deception to forage with wasps and bees at flowers near the rufous-tailed and steely-vented hummers. Territory parasitism is a strategy also used by Cuban bee and Mexican bumble bee hummingbirds. They are usually able to sneak their fill without being discovered.

In an hour or so, when the steely-vented hummer is satisfied, it departs. The coffee flowers produce a bit more nectar during the afternoon, so the shy but persistent fork-tailed emerald returns periodically to glean them.

Aggressive hummingbirds, such as the rufous-tailed in the tropics or the blue-throated, Allen's, and rufous in the United States, do not defend flowers unless they are richly rewarded with nectar. When production wanes, they search elsewhere, leaving less desirable flowers to meeker birds.

Hummingbirds fuel their rapid metabolism by feeding frequently throughout the day. They do not feed at night. In order to conserve energy, hummingbirds have the ability to enter torpor, a short-term form of hibernation. Respiration, heart rate, and general metabolism slow considerably. Body temperature drops from an average of 105 degrees Fahrenheit to as low as a few degrees above ambient air temperature. (In the South American Andes, where overnight temperatures plunge below freezing, the Andean hillstar seeks sheltered roosts in the mouths of caves or mine shafts.) Normally torpor occurs at night. During unexpected cold snaps, hummingbirds may remain in torpor during the day.

In one study, an Anna's hummingbird that partook of normal feeding, preening, and territorial defense, then slept regularly during the night, consumed 10,300 calories. For an equal period, an Anna's that was active by day but entered torpor at night used only 7,600 calories. Torpor requires as little as a fiftieth of waking energy levels and a fifth of sleeping energy. Energy consumption of a hovering hummingbird compared to a torpid hummingbird equals a human running at 90 mph versus a human walking briskly.

There are dangers in becoming torpid. Hummers may chirp feebly if disturbed but they cannot flee from predators. If knocked from a perch, they are not able to regrip another twig. Emerging from torpor requires energy to warm organs and tissues back to functional temperatures. Somehow a torpid bird must monitor its energy level and recover from the zombielike state before its reserves are completely exhausted.

In normal sleep, hummingbird internal temperatures drop four to eight degrees. If air temperatures drop lower than about 95 degrees Fahrenheit, a bird may enter torpor, although this reaction depends as much upon the individual's fat reserves and general health as air temperature.

Two hummers of the same species might roost in the same tree on the same night. One may enter torpor and the other may not. Anna's hummers have been observed eating ravenously before dark. If they store enough food in the crop, they may simply sleep through the night. Rufous hummers in the same region may not feed as heartily at dusk. They would be more likely to go into torpor. Generally, tropical hummers do not stay in torpor as long or drop to as low an internal temperature as northern or high-altitude species.

If going into torpor, a hummingbird will erect its feathers away from its body to allow heat to escape. Breathing slows. The waking pulse of a resting blue-throated hummingbird is about 480 beats per minute. In torpor the heart beats only thirty-six times per minute. A study of rufous hummingbirds showed they might enter torpor any time from dusk to midnight. Awakening, however, occurred within two and one-half hours of daylight.

Waking Allen's and Anna's hummers increase body warmth at one to two degrees per minute.

They can fly when internal temperatures reach 86 degrees Fahrenheit. Larger hummers take slightly longer to warm than smaller ones. In Brazil, hummingbirds have been observed remaining in torpor until well after sunrise, when they are heated by solar radiation.

The ability of hummingbirds to enter and exit this deathlike state led to early legends about their supernatural powers. Bernardo de Sahagun, a Franciscan monk in Mexico in the mid-1500s, wrote of marvelous *pajaros resucitados*, birds that could die in winter, then resurrect. The Aztec name *vitzili* has similar meaning. Early Spanish explorers falsely believed hummingbirds impaled their beaks in tree bark and died during the dry season, then were revived as tropical rains fell.

An acquaintance told me of an outing with his grandfather some years ago. The sharp-eyed lad spotted a hummingbird hanging upside down on a wire beneath a canopy on a storefront. His grandfather took the supposedly dead bird down for the boy to admire, then stuck it in his shirt pocket and continued with his errands.

Later, as they were driving home, the old man felt a strange palpitation near his heart. He swerved his car to the side of the road just as the formerly torpid hummer burst from his pocket. Muttering something about miracles, the incredulous old fellow rolled the window down and watched the resurrected bird fly away.

Risks and Rewards of Promiscuity

Birds that mate for life, such as magpies and ravens, often have modest courtship rituals. Species with sexually promiscuous males—prairie chickens, grouse, and birds of paradise, for example—have elaborate sounds, displays, or plumage to bond males and females during the short mating period. Hummingbirds have casual sex but intricate courtships.

Most male hummingbirds breed with multiple females each season. Some females accept more than one mate. Male hummers, who have no part in nesting or chick-rearing except for the act of copulation, perform dazzling displays to attract females. In open areas, sun shimmering from a male's crest or gorget reflects light like a flashing theater marquee. Swift aerial dives add to the show. For hummingbirds of dense tropical forests, song becomes important.

As might be expected of creatures dependent upon high-energy foods, a male hummingbird's initial territory selection is based on the area's nectar resources. Migrant males arrive at their summer locations two to three weeks ahead of females. They claim ownership of wildflowers by hovering face-to-face, then rising high in the air while trying to intimidate one another with aerial skills. Tails flair in such confrontations, making hummers look larger or revealing striking feather patterns.

Sometimes combatants collide in mid-air or fall to the ground for a brief tussle. Bills are seldom used in such struggles because they are fragile and could break. In rare instances, however, intense fights have resulted in one hummer being poked or knocked unconscious by another.

A few general rules apply to territoriality. An owner will be more aggressive if his territory contains abundant food to replenish energy spent in defense than if nectar is scarce. Large or brightly colored hummers are more effective at warding off intruders. Smaller or dull-hued birds are better suited to pilfering nectar.

Hummers do not limit territorial defense to actions against other hummingbirds. Keeping watch from a perch that overlooks wildflowers or a feeder, they will buzz butterflies, bees, cats, humans, and other birds including orioles, crows, and hawks. In one instance, a tropical hummingbird was observed joining larger honeycreepers and trogons to mob an intruding owl.

Courtship is difficult to study in such small, mobile, promiscuous birds. Migrant female hummingbirds probably select a suitable nest site with nearby sources of nectar soon after they arrive on the summering grounds. Nest construction is almost finished before female hummers seek the services of males.

When a female enters the territory of a potential suitor, she may be looking for nectar. He may treat her aggressively. Her reaction, however, is unlike that of a rival male, who would fight or retreat. The female perches and may eventually be lured or intimidated into mating with the male.

In open areas, hummingbirds such as Anna's, Allen's, or Costa's use territorial defense flights in

Before mating, a male calliope hummingbird performs a buzzing, U-shaped aerial display. After his brief copulation with the female, the male has no further parental duties. (Photo copyright by Charles W. Schwartz)

Above: In Central America, male white-bellied emeralds gather on leks, where they repeat a monotonous, chipping chorus to entice females to mate. (Photo copyright by Carol Farneti)

Left: During their early spring breeding season, male Anna's hummingbirds sing from conspicuous perches. Their iridescent crown and gorget feathers reflect the sun in a shimmering beacon of light. (Photo copyright by Hugh P. Smith, Jr.)

which males hover as much as a hundred feet high. They swoosh down in U- or J-patterned dives, repeating the flights several times. Ruby-throats also dive to display, but use a lower track, swaying broadly like a pocket watch dangling from a chain. Males also employ shuttle nuptial displays. A male faces a perched female and repeatedly flies back and forth in a horizontal arc in front of her. He may sing one or two squeaky notes or make noises with vibrating feathers. She may spread her tail in recognition. Occasionally females shuttle for males. The flights usually end in copulation.

Hybrid offspring between species such as Anna's, black-chinned, and Costa's have been documented, especially on the fringes of their ranges. Biologically, hybrids are a waste of reproductive effort because they are usually sterile and cannot pass traits from parents to succeeding generations. Males have no trouble recognizing the gaudy plumage of other males. But they may mistake a similar female for one of their own kind and display for her.

When several species of hummingbirds in which females look alike occupy a certain range, males are likely to perform complex displays. Her consent to breed is triggered only by a specific flight—the J-pattern with popping noises for an Anna's versus a low U-shaped dive for the black-chin, for example—so she should not yield to a male of another species.

Gary Stiles, who studied Anna's hummingbirds, has acquired some of the most complete information on courtship of North American hummers. Anna's breed as early as December, much sooner than other North American species. They are sustained at this time by manzanita and currant shrubs blooming in the warm Southwest. Males will have recently finished their annual molt and will be feathered in dapper rose and green iridescent plumage.

At the beginning of the breeding season, male Anna's choose territories of about a quarter acre each. From a conspicuous perch, the male chatters a scratchy, repetitive song and sways side to side. He aligns himself facing the sun so any intruder will see light reflecting as he expands the elongated feathers of his gorget.

Male Anna's claim territory by circling in flight above the boundaries. Threat or courtship diving displays are enacted most frequently on sunny days, always oriented to the sun. In these performances, the male hovers at six to twelve feet, then rises sixty to one hundred feet, all the while peering at a rival male or prospective mate. The male power dives straight toward the intruder. Very near the interloper, the male pulls up in a vertical arc. The change of direction results in a pop of air rushing across the tail. The male hovers, staring at the intruder and chattering. He may repeat the whole process several times.

When a female Anna's enters his territory, a male initially treats her as a rival. If she reacts passively, he will perform a courtship dive. She may show interest by ruffling her crown and gorget feathers, spreading her tail, bobbing her head, and chattering. If she flies, he will chase her. Females are thought to lead suitors toward their nests to reduce the aggression males generate within their own territories. If the female perches amid vegetation, the male will perform his horizontal shuttle flight. Then they mate.

Stiles writes of "copulation with the female perched, male hovering over and descending onto her back, where he clings, half perched, half hovering; cloacal contact lasts about two seconds." Afterward, she preens and goes about her nest-building routine. He returns to his territory.

Although northern hummingbirds rely primarily on dives and shuttle flights to forge nuptial bonds, males of a few tropical and Caribbean species are garbed in plumage specially adapted for breeding rituals. Male Jamaican streamertails use the sound of their long black tail feathers to attract attention. South American red-tailed comets, green-tailed trainbearers, and violet-tailed sylphs flash long, colorful tails at mates.

The marvellous spatuletail of Peru possesses four main tail feathers. The inner two are long, slender, and pointed. The outer pair have bare, curving shafts for most of their length, but they end in rounded, purple-sheened black tufts. In courtship, the male hummer raises the unusual feathers to frame his face. Racket-tailed coquettes and booted racket-tails, both from northern South

America, have pendulum-tipped tail feathers on straight shafts. Male tufted coquettes, horned sungems, and rainbow-bearded thornbills are a few of the tropical species adorned by colorful feather headdresses.

Hummingbirds that dwell in forests have less elaborate courtship flights than those of open areas. Generally, the feathers of woodland hummingbirds are more drab than those of meadow-dwellers. Many hermits, robed in shades of green, gray, and brown, rely on sound rather than sight to attract mates.

Most birds have fairly large sound-producing muscles extending from the windpipe to the breastbone. Hummingbirds do not. They possess two small sets of vocal muscles in the trachea, but the noises produced are limited to high-pitched, soft twitterings. These quiet notes have the same function as songs of warblers and vireos—advertising territories and sexual partnership—but hummingbird voices may not carry as far.

Especially in the tropics, a few to several dozen male hummingbirds of the same species will band together to sing. They gather on "leks," traditional songfest sites, each covering about half an acre. Leks are used year after year. Youngsters learn rhythmic patterns from elders, each of whom has his own guarded singing perch.

Blue-chested hummingbirds, native to Costa Rica, sing in the morning and late afternoon during the mating season, using the rest of the day to feed, preen, and interact with other hummingbirds. Rufous-tailed hummers and green violet-ears, also found in Central America, sing throughout the day. They stop only briefly to feed or chase away territorial invaders. Little hermits in Trinidad have been documented repeating the same simple call at two-second intervals about twelve thousand times per day. Their mating season lasts from November through July.

Long-tailed and little hermits, in which males and females look alike, employ leks throughout Central and South America. Reddish hermits use

Sun shimmering on the gorget of a male ruby-throated hummingbird sends a message of territorial ownership. It may also invite receptive females to mate. (Photo copyright by Luke Wade)

Above: Male broad-tailed hummingbirds aggressively de-fend their territories against other hummingbirds, bees, butterflies, and sometimes larger birds. Hikers in the Rocky Mountains often report face-to-face confrontations with these inquisitive hummers. (Photo copyright by Sid Rucker)

Left: When establishing their territories, hummingbirds often hover in front of each other or flare their tails to reveal striking feather patterns. (Photo copyright by Hugh P. Smith, Jr.)

leks only in parts of their range. Planalto hermits, native to Brazil, do not use singing grounds at all.

Since there is little difference in plumage between male and female lek-using hummingbirds, breeding strategies differ from the gaudy power displays of open-area hummers. Once the chorus of singing males draws a female onto the lek, she signals her readiness to mate by sitting quietly and attentively. (Passive male birds are occasionally, accidentally, courted by other males.)

Exact breeding rituals vary from species to species. The South American male rufous-breasted hermit nears a waiting female and flicks out its long white tongue. She chortles a soft, sweet song if she is receptive. The little hermit male, plumed in black facial feathers like a masked Zorro, will dart and dip in the air only inches from his sweetheart. Buzzing his wings more loudly than normal, his nuptial flight may last ten minutes or more. All the while, the placid female's head follows him back and forth as though she were watching a fencing duel.

Many woodland hermits are so well camouflaged that they blend with the vegetation of the lek unless they bob their tails as they sing. Most have light spots or bands on the tail. Some also have white streaks above or below the eye that show up as they move about in the dense underbrush. More colorful songsters, such as the green violet-ear, blue-throated goldentail, and white-eared hummer, move higher in trees to sing and be seen.

After chortling males have attracted females to the lek and performed their introductory flights, females probably return to nesting territories with males in pursuit. This behavior has not been well-documented because it is so hard for observers to follow small, rapidly flying birds through dense underbrush. It is believed, however, that in at least some species mating takes place away from the lek.

It is quite unusual to see male and female hummingbirds together peacefully for more than a few moments. Their promiscuous breeding behavior does, however, have an adaptive advantage. One male may pass traits to numerous offspring in a relatively short time. Thus genetics of the species change rapidly. Features that enhance a bird's chances of success, such as bill shapes or feather color, allow diversification into various terrains.

Hummingbirds have successfully adapted to a wide range of habitats. They represent an extreme in high-risk, high-reward lifestyles. Differences in male and female roles may be responsible in a large part for this success.

A Lichen-Covered Hideaway

As breeding season approaches, hormones urge a female hummingbird to gather nesting materials. Inside her body, minute ova, or eggs, ripen in the left ovary. In most types of birds, only the left ovary develops normally. The other shrivels to reduce the female's weight in flight. Similar hormonal changes activate and enlarge the male hummer's testes to many times their off-season weight.

Millions of sperm, deposited during the brief mating tryst, swim up the female's oviduct. In most female hummers, two ova are fertilized. Cells differentiate into yolk and albumen and multiply rapidly. As the eggs pass into the lower oviduct, thin layers of shell are deposited around them.

Nests of each type of hummingbird are recognizable by their unique designs. Those found in the United States and Canada are cup shaped, well camouflaged, and average two inches across. Beginning with a small core that is anchored to a branch, they are laced together with spider webbing. The interior is lined with plant down, animal fur, soft grass, or moss. The outside is decorated with bits of lichen, bud scales, or bark. Nests are often overlooked because they resemble knots on a tree.

While birding in Cape May, New Jersey, one spring, I found a female ruby-throated hummingbird building a nest about fifteen feet up on a branch of a walnut tree that drooped over a quiet gravel road. I would not have discovered her hideaway except I was watching a warbler through binoculars and the hummingbird landed in the foreground. Then I observed the hummer go to the same branch three times in a row.

She had just begun the nest, so I could only discern a few bits of lichen piled there. Each time she came back, she hovered or sat, using her beak to poke bits of vegetation. When she turned at one perfect angle, I could see sunlight glistening on strands of spiderweb in her beak.

I watched her progress over the next three days and saw the bowl of the nest grow. The dainty hummer formed a depression by pushing her breast against the walls of the nest. She appeared to use her feet to pack fibers into the floor. The cup was sized to seal against her belly, insulating eggs or hatchlings inside. As the young grew, the interwoven design would stretch to accommodate their increasing size.

Female hummingbirds take from a day (in species that renovate nests from previous seasons) to two weeks to fashion their nests. They will steal materials from other hummers in the area if available. Some continue building edges higher or adding decorations to the outside after they have laid eggs.

Many hummingbirds nest near rivers and streams or on branches overhanging roads. Blue-throated hummingbirds, which occupy cliffs in streamside habitats, recycle old nests. Outside dimensions are smaller than a baseball. The head and tail of the incubating female extend over both sides. Ornithologist Herbert Brandt dissected a

Photographer Clayton Fogle discovered an Allen's hummingbird nesting on a dusty chain link in an old shed. He returned several years in a row to find what was probably the same female nesting at the same site. One season she had babies in this nest while also tending another nest in a chain link about fifteen feet away. (Photo copyright by Clayton A. Fogle)

Left: Broad-tailed hummingbirds often choose nest sites sheltered by overhanging branches. Photographers Kent and Donna Dannen encountered this incubating female after a hailstorm in the Rocky Mountains. (Photo copyright by Kent and Donna Dannen)

Above: At about two and one-half weeks of age, hummingbird chicks are fully feathered. This young black-chin will test its wings by anchoring a foot on the nest and fluttering rapidly. (Photo copyright by Luke and Margie Wade)

blue-throat nest attached to a cabin along Ramsey Creek in southeastern Arizona. It had been used for ten years and was more than twice the height of normal nests. Brandt estimated it contained nearly fifteen thousand miles of spider webbing.

Broad-tailed hummingbirds frequently choose the rough bark of a pine, spruce, or fir tree on which to anchor their nests. They look for sheltered sites under an overhanging branch that will keep heat from radiating into the cold night air and will deflect rain or snow flurries common in their mountain habitat. Calliope hummingbirds, which also nest in the mountains, often orient their nests facing east so they will be warmed by the morning sun. They build near knots or cones for camouflage. Rufous hummingbirds choose sites low in conifers for the first nests of the season. Nearer the ground, temperatures are more constant in spring. By summer, if rufous hummers nest a second time, they choose sites higher in broadleaf trees.

Ornithologist Bill Calder has fitted hummingbird nests with temperature sensors hidden in false eggs. He discovered the structures are so well insulated that a brooding female can keep eggs at 95 to 97 degrees Fahrenheit even when temperatures outside the nests drop into the forties.

Tropical hummingbirds have nesting considerations other than temperature regulation. Streamertails nesting on the rainy side of Jamaica construct their nests so loosely that moisture drains through them. Several species of hermits time nesting to coincide with abundant floral displays, which occur during the rainy season. To protect their nests, they attach them to the undersides of wide palm fronds that serve as living umbrellas.

Hummingbird nests come in various shapes and sizes. The nest of the bee hummer is the world's smallest—only an inch across and half an inch deep. A nickel, if laid inside, would barely fit in the egg cup. The vervain hummer, only slightly larger than the bee, has a nest with an egg cup about the size of a quarter. The Venezuelan sooty-capped hermit suspends its nest on a cable of spider webbing. To keep it from tipping and spilling the precious eggs—since the cable attaches to only one side of the fist-sized nest, the female lashes small balls of dried mud below. The weight of the mud counterbalances the nest and keeps it upright.

Many homes in the tropics do not have screened windows, thus hummingbirds can fly inside to search for insects and cobwebs. It is not unusual for tropical hummers to nest inside in the midst of human activity. Their presence is welcomed by most families. Female hummingbirds accustomed to people have plucked hair from the heads of surprised humans to weave into nests.

Nest construction and chick-rearing put great strain on a female hummingbird. She may choose a nest site because of abundant flowers and defend the area briefly. But as she builds the nest, incubates eggs, and feeds chicks, there is little time for guard duty. She may be forced to pilfer nectar if another bird usurps her territory. Females attack songbirds, cats, squirrels, and other intruders that approach the nest.

Eggs are laid shortly after the female completes the nursery. Most hummingbirds lay two eggs, but the giant hummingbird of western South America produces only one. The first is normally deposited in the morning. The second translucent pinkish egg arrives two days later. As they are exposed to the air, eggs become opaque white. Some species begin incubating after the first egg, and their young hatch at two-day intervals. Others wait to incubate until the second egg appears. Their chicks hatch simultaneously.

Compared to other birds, which lay eggs equaling 2 to 4 percent of their body weight, hummingbirds lay remarkable eggs, comprising 10 to 20 percent of the mother's mass. In human terms, this would compare to a 130-pound mother bearing a twenty-pound child. Even so, hummingbird eggs are tiny. Average size is less than half an inch long, weighing less than 0.02 ounce. Theoretically, one could mail five dozen hummingbird eggs with a first class postage stamp. Hummingbird eggs are elliptical rather than pointed at the end. In size, shape, and color, they resemble navy beans. The largest hummingbird egg, that of the giant hummer, approximates the size of a plump raisin.

Incubation requires fifteen to twenty days, depending upon the species. Hummingbirds spread their feathers to make body contact with the eggs.

Many of the tropical hermits reproduce during the rainy season, when flowers are abundant. They build nests under palm fronds, which function as natural umbrellas. (Photo copyright by Gerard Lemmo)

The tight fit of the nest rim against the mother and the insulative qualities of nest construction maintain high incubation temperatures. Female hummingbirds sit tight in cold, rainy periods and do not enter torpor at night while incubating, as drops in temperature delay embryo development. Females leave the nest for short periods to forage but keep the eggs covered 60 to 80 percent of the time.

When hummingbird chicks are ready to meet the world, they use a sharp appendage on their beaks to break out of the shell. The egg tooth falls off soon after hatching. Young hummers are smaller than a wad of chewed gum at birth. They have no feathers and their eyes will not open for nearly two weeks. They consist mostly of a bald head attached to a digestive system.

After hatching, chicks immediately receive a nutritious broth of nectar and insects. Meals the first few days may so completely fill them that tiny crops bulge from the right sides of the babies' necks like tumors. At birth, broad-tailed hummer chicks weigh about 0.012 ounce. It would take 1,332 newly hatched chicks to make a pound.

Unlike other species, hummingbird chicks are not covered in down. Instead, two rows of pin feathers sprout from their backs. The sound of mother's wings and the rush of air across developing feathers stimulates young birds to gape. Bright mouth linings provide a target for the female. Mother hummingbirds dispense food equally to both chicks on each feeding visit.

Hummingbird nests are loosely woven from plant fibers and spiderwebs. They stretch with the babies as they grow. (Photo copyright by Clayton A. Fogle)

Some years ago ornithologist William Finley described a female rufous hummer feeding her young:

After she had spread her tail like a flicker to brace herself, she craned her neck and drew her dagger-like bill straight to the hilt and started a series of gestures that seemed to puncture him to the toes. Then she stabbed the other twin till it made me shudder. She was only giving them dinner after the usual hummingbird method of regurgitation, but it looked to me like the murder of the infants.

Since deaths of youngsters occur mostly during cold, wet weather, females brood babies eight to twelve days until they begin to attain feathers and regulate their own internal heat. Mother will leave briefly, twenty to sixty times a day, to find nectar. She may also catch as many as two thousand insects per day if they are available, since chicks need large amounts of protein to grow and develop. After feeding nestlings three or four times, mother will forage for herself a few minutes. One female broad-tail in Colorado was seen leaving the nest for her first feeding fifteen minutes before sunrise. The last foray occurred at sunset.

Mother gathers the droppings of newborns with her beak and removes them from the nest. As youngsters become more mobile, they instinctively back to the edge of the cup and squirt fecal material over the rim.

By two and one-half weeks old, baby hummers are covered with feathers, their bills have elongated, and they can groom themselves. They test their wings by anchoring a toe in the nest and fluttering rapidly. After three weeks, they leave the nest, usually on a calm morning. Fledglings have good mid-air control but often botch the first few landings.

Well-fed chicks are slightly plumper than their mother when they depart. Mother continues to feed them for another two to four weeks, with youngsters waiting on a perch rather than follow-

Lichen-covered nests, such as the one being built by this female ruby-throated hummingbird, are frequently overlooked because they resemble knots on tree branches. (Photo copyright by Steve and Dave Maslowski)

Above: Nests of each species of hummingbird are unique in design. The interior of a berylline hummingbird's nest is lined with down from the Arizona sycamore tree and the outside is decorated with bits of lichen. (Photo copyright by Connie Toops)

Left: Although hummingbirds have few natural enemies, they are occasionally nabbed by praying mantises that lie in wait on flowers or feeders. This white-bellied emerald was captured by a mantis in Belize. (Photo copyright by Carol Farneti)

ing her in flight. Fledglings also explore, learning to recognize nectar-bearing flowers and mastering the art of hovering while drinking nectar or catching insects. Siblings chase each other, gaining skills they will later need to defend territories.

Some hummingbirds nest twice during the season. A female may begin refurbishing or constructing a second nest while she is still caring for young of the first brood. The nest just vacated, which may be infested with lice or other parasites, is seldom used again the same season.

As many as half the hummingbirds hatched during the year do not survive to adulthood. Vagaries of weather, poor food supplies, and predation combine for low reproductive success. Losses are offset by fairly long lives for birds that reach adulthood. Life spans average about five years.

In June 1976 ornithologist Nick Waser placed band X18025 on a female broad-tailed hummingbird that was at least one year old. She was recaptured several times, the last being in 1987 at the age of at least twelve. After the banding season of 1991, Bill Calder reported recaptures of four other broad-tails who were eight years old. Jim Johnson and Marguerite Baumgartner each documented nine-year-old ruby-throats, and Elly Jones recaptured a six-year-old calliope originally banded by Bill Calder. A distinctive wild blue-throated hummer was observed over the course of twelve years. Several zoos and aviaries have raised hummingbirds ten to eleven years old. One captive female planalto hermit in Brazil reached age fourteen.

Very few predators are agile enough to catch hummingbirds in flight. Merlins, kestrels, tiny hawks, and bat falcons—all small, rapidly flying hawks—occasionally nab them, as do large tropical flycatchers. Roadrunners attack hummers while they are perched or hovering, and orioles, who compete with them for nectar, have also killed hummingbirds.

Incredible as it may seem, invertebrates are also predators. Now and then hapless hummers get stuck in spider webs, especially strong nets of large tropical species. A ruby-throat was caught and wrestled to the ground by a large dragonfly as stunned birders in the East looked on. Preying mantises lie in wait on flowers and feeders, then strike with lightning speed. On rare occasions when hummers dip too near the water, they may be gulped by large frogs or fish. Lizards, snakes, squirrels, and jays prey on hummer eggs and chicks.

For large predators, it is doubtful a hummingbird provides much nutritional reward. Certainly none of these creatures are in the habit of making regular meals of hummingbirds. Perhaps it is because hummingbirds have so few natural enemies that they are seemingly fearless around humans and other large animals.

The Longest Journey

It is mid-July in the Adirondack Mountains of northern New York. A young ruby-throated hummingbird teeters on the edge of a lichen-covered nest that has been home for the past sixteen days. With a burst of energy, it lifts into the air, landing clumsily on another branch of its hophornbeam tree a few yards away. This is the first of many flights in which the fledgling will explore along the stream trickling near its birthplace. Gradually it will learn to find nectar at red bee-balm and fragrant honeysuckle flowers.

During the next two months the fledgling gains strength and agility. No longer dependent upon its mother for food or protection, it eats at every opportunity. Compact, fat-laden reserves of energy build in its tissues. As day lengths fade, the little bird feels a deep, hormonal urge. One crisp morning in early September the wind is from the north. The young bird makes its rounds of wildflowers in the streamside meadow, then heads south into unexplored territory.

Flying above the treetops, the young hummingbird follows contours of the mountainous terrain. With the wind at its back, miles pass quickly. By mid-afternoon it spies a meadow similar to the one it left behind. It flits among bee-balms, lobelias, and dragonhead flowers, ravenously lapping their nectar.

Fair skies and northerly breezes hold for the next two days. The fledgling continues in a south-southwesterly direction after each morning's meals. It flies alone, a few feet above the vegetation. By

mid-afternoon of the third day, it reaches northern Virginia.

The weather has taken a turn for the worse. Clouds hide the surrounding Blue Ridge mountain peaks. Chilly rain squalls sweep across the narrow valley where the hummer has landed. Between showers, the little bird discovers a few nectar-producing flowers. As darkness falls, it huddles on a branch of a bushy red cedar and fluffs its feathers against the chilly dampness.

The next morning the wayfaring hummer explores the valley and discovers a flower garden behind a rural home. It drinks from a variety of blossoms—red-hot poker, salvia, nasturtium, gladiolus. Then it sees a red container dangling from a hook at the back of the house. The curious bird investigates and is rewarded with sweet, free-flowing nectar. For two more days the hummingbird remains, refreshing itself at the flowers and feeder until the weather clears.

Finally the front passes and the ruby-throat continues its migration. By feeding heartily each morning and evening, it is able to maintain high energy levels. Flying during the middle of the day, it progresses steadily southward while fair weather prevails. In less than a week it parallels the length of the Blue Ridge and Great Smoky mountains.

Coursing above the treetops in northern Alabama's foothills, the ruby-throat hears soft chittering. It descends through the forest canopy to a creek meandering between oaks and long-needled pines. Growing along the sandy stream banks are

First-year hummingbirds, including the ruby-throated, migrate alone rather than with parents or experienced birds. They are born with an innate sense of where and when to fly. (Photo copyright by Connie Toops)

thousands of waist-high jewelweeds, their red-orange flowers dangling like Japanese lanterns from the branches. Feeding at these blossoms are scores of ruby-throats—adult males with red gorgets flashing, green-backed females, many drab birds like himself. A few adult males chase the youngsters, but nectar is abundant and most of the hummers feed with a determined sense of purpose.

The young traveler stays four days, feasting on jewelweed nectar. Then it heads south again, following lazy, cypress-lined creeks. Along their banks the hummer finds occasional rest stops where jewelweed and stately cardinal flowers grow. In another three days it reaches the edge of a huge green-gray body of water that has no visible shore on the opposite side.

A breeze from the southeast adds a touch of white foam to the incoming waves. The hummer retreats inland a short distance to a field overgrown with brush and trumpet creeper vines. The little bird, who has now flown nearly 1,100 miles (twenty-three million body lengths) since leaving home in upstate New York, rests and waits.

Days are sultry, filled with humidity rolling in from the tepid Gulf. Less than a week later, however, the hummingbird detects a change. A line of showers passes. The air behind them is brisk; the horizon over the water is no longer obscured by haze. By morning, steady wind blows from the northwest. The hummingbird lifts from its perch, and riding a tailwind, begins the longest single journey of its life. Skimming low over the water, it sets a course almost due south.

Hours pass as the sun arcs above the little bird. Tiny wings move up and down three thousand times each minute. In eleven hours of daylight since leaving the coast, the hummer has covered 275 miles, half the width of the Gulf of Mexico. As dusk settles over the vast body of water, there is no perch to rest upon, there are no flowers offering nectar refreshment. The weary bird flies into the darkness.

Streaks of red appear on the horizon the fol-

Before migration, hummingbirds feed heavily to accumulate extra fat that will fuel their long journeys. Using these reserves, ruby-throated hummingbirds fly non-stop across the Gulf of Mexico. (Photo copyright by Luke Wade)

Residential patterns of a few species have been influenced by increased urban landscaping. Anna's hummingbirds have expanded their range east into suburban areas of Arizona and north into populated parts of western Oregon and Washington. (Photo copyright by Hugh P. Smith, Jr.)

lowing morning as the three-inch bird strokes its four-millionth wingbeat since leaving the coast. The tailwind gave way to still air during the night, slowing the hummer's momentum slightly. Shortly after sunrise, however, the hummingbird detects a thin line of white ahead. It is the north shore of Mexico's Yucatan Peninsula. Exhausted, the hummer drops into the mangrove forest behind the beach and sits on a branch, hidden among the thick green leaves.

After a well-deserved nap, the young bird is famished. It spies a pineapple-shaped growth with red at the top on a nearby tree. The traveler guzzles life-renewing nectar from these bromeliad flowers and catches a few gnats around the cup-shaped base of the plant. It alternates resting and eating for another few days, recovering from the trans-Gulf journey.

The urge to head south still dominates the hummer's intentions. Reverting to its feed in the morning, fly in the afternoon, feed before dusk strategy, the hummer spends several more days winging south above fields and woodlands of the Yucatan. Below, patches of forest have been cleared for pastures and sugarcane. Acrid odors of burning slash hang heavy in the air.

Finally the frenzy to push south mitigates. The ruby-throat selects a section of woodland along the Belize–Guatemala border that will be home for the next five months.

This ruby-throat was lucky. Half to three-quarters of the offspring produced during the summer nesting season do not reach adulthood. Migration is the most stressful period in a nonsedentary bird's life. Migrant hummers face the vagaries of weather and the threat of predation en route, but also, with each passing year, they lose habitat to human development, reducing available rest stops and suitable overwintering areas.

Why, then, would a bird risk migration? For many species, staying year-round in the tropics would overcrowd limited food supplies and nesting habitat in spring and summer. Adventurous migrants exploit seasonal food abundance and

The black-chinned hummingbird is one of several species that risk a long migratory flight to reach summer habitats offering abundant food and nesting areas. (Photo copyright by Luke Wade)

lower population densities of outlying habitats. All hummingbirds of the United States and Canada, with the exception of Anna's in coastal California and a small population of Allen's on California's Channel Islands, are migrants. Of the entire hummingbird family, however, migrants comprise a small percentage.

There is some incredible mechanism—a function we do not yet fully understand—packed into the very small space of a migrant hummingbird's brain. It is a more marvelous miniaturization than the silicon chips that store and access information in computers. Young migrants, such as ruby-throats and rufous hummingbirds, are born with an innate sense that tells them when to fly, what route to take, how heavily to feed, where to rest, and when to stop.

Young birds do not fly with their parents, nor do hummers form protective flocks, as migrant shorebirds do. Despite folktales, they do not hitch rides on the backs of migrating geese or swans. But amazingly, hummingbirds navigate back to

their summer territory with such accuracy that they will return to the same patch of perennial wildflowers or hover at the nail where a favorite feeder was suspended, even if food has not yet been put out by the human provider. Acquaintances who cage-reared an injured black-chinned hummer told me the captive sensed the proper time for fall migration. Even though it had but one good wing, the black-chin tried in vain to fly south no matter which direction they oriented the cage.

How is navigation, which for a rufous hummingbird may mean covering as many as 2,500 miles, accomplished? That question has not been answered completely. The journey of the young ruby-throat mentioned above was developed from composite information rather than by following one bird. So far, the only individual recaptured after migrating to Mexico is an adult male black-chin banded at Sonoita, Arizona, in July 1988. It was netted again a thousand miles south in Jalisco, Mexico.

Ornithologists know some migrant songbirds

Hummingbirds from the United States and Canada spend their winters in Mexico and Central America. Loss of rest stops and overwintering habitat, such as this forest in Belize, will adversely affect their populations. (Photo copyright by Connie Toops)

Flying fast and low over the waves, ruby-throated hummingbirds cross the Gulf of Mexico each spring and fall. They sense the arrival of fronts that bring weather favorable for migration. (Photo copyright by Connie Toops)

use the sun and stars as a compass and have internal clocks that adjust for the movement of these celestial bodies. Homing pigeons detect familiar odors and use them to navigate back to the loft. They also hear ultra-low sounds, orienting to low-frequency wave noise from oceans as far as a thousand miles away.

Bees discern polarized light patterns in the sky. Sharks detect and navigate by minute changes in electromagnetic fields, which vary with distance from the north pole. Birds may use these senses as well, but organs to interpret magnetism and polarized light have not yet been identified within their anatomies. Birds are known to sense approaching cold fronts, which bring favorable wind shifts, by changes in air pressure in their middle ears.

Studies of migrant birds in Europe revealed instructions for migration are inherited. Young starlings were intercepted in the midst of their first migration. They were transported and released at a new location. They continued on the original heading, arriving at a nontraditional wintering site because their innate urges did not compensate for translocation in the midst of the journey. In a similar test, migrating adult starlings were also captured and transported to a new location. When released, the experienced birds relied on knowledge of the proper route from a previous trip. They took a new compass bearing and arrived at their normal overwintering site.

Hummingbirds are probably governed by similar genetic programming. First-year birds instinctively fly in a particular direction for a certain distance. Young birds seem to recognize the area where they were born and the area where they overwinter. In subsequent years they retrace the migratory route between these areas using knowledge learned on the first trip.

Rufous hummingbirds, for example, normally migrate in a corridor that stretches from southern Alaska to central Mexcio, bounded by the Rocky Mountains on the east and the Pacific Ocean on the west. Fast-moving autumn storms with strong westerly winds may blow migrant hummers far off course. Adults compensate, changing their route to reach Mexican wintering areas. Young birds fly along pre-programmed vectors

after the storm, usually halting hundreds of miles out of their normal range when they reach a coast.

Nonnative landscape plants that provide nectar throughout the winter and humans who keep hummingbird feeders stocked year-round may keep some of these disoriented hummers alive. Rufous hummingbirds banded as immatures have returned to the same out-of-range wintering spots in succeeding years, indicating they learned the aberrant route.

The effect of feeders on migrant hummingbirds is a subject of contention. A few people think it is best to take hummingbird feeders down on Labor Day to urge hummers to head south. Every hummingbird expert I know disagrees. Says Nancy Newfield, who has studied hummingbird distribution and behavior since 1975, "There is no magic date to take feeders down. Food alone does not hold hummingbirds. If it did, none would leave our summer feeders or the abundant winter flowers of the tropics."

Since 1979 Nancy has banded more than 3,500 hummingbirds at her home near New Orleans and in adjacent areas of Louisiana, Texas, and southern Mississippi. She notes many hummingbirds migrate south long before food sources are exhausted, perhaps even timing migrations so favorite blossoms, such as cardinal flower, jewelweed, or Turk's cap will be in peak bloom periods.

Nancy has determined most hummers lingering at feeders in the fall are underweight. Healthy ruby-throated hummingbirds en route south in the fall weigh in at Newfield's banding station at 0.175 ounce or more. Spring weights for the same birds would average slightly over 0.1 ounce. The extra fall fat will be metabolized during the long flight across the Gulf or around the coastal bend. Healthy, heavy birds do not remain at Nancy's feeders long before continuing their journeys. Leaving feeders up throughout the migratory period may increase the survival chances of late-hatching hummers and underweight birds that did not find abundant flower nectar in the summer.

Along coasts and in mild areas where cold

Although the migration routes of rufous hummingbirds coincide with abundant wildflower blooms, feeders provide a welcome additional source of nectar. (Photo copyright by Connie Toops)

snaps are temporary, a few nonmigrating hummingbirds may survive the winter by using feeders and finding local flowers in bloom. These birds are not held from normal migration by feeders or landscaping. Instead, they may have a nutritional deficiency or an injury that keeps them from migrating. Feeders or flowers maintained for them may keep them alive.

Residential patterns of a few hummingbirds may be changing in response to landscaping in expanding urban areas. Buff-bellied hummingbirds are being reported more frequently along the suburban Texas coast. Only a few years ago it was unusual to spot them north of Corpus Christi. Several have recently wintered near ornamental plantings in Houston and coastal Louisiana.

The first Anna's hummingbird nest was discovered in the Tucson area in 1964. They are now common in suburban areas of Arizona, especially in fall and winter. Anna's hummers nest there in fair numbers, usually in eucalyptus or ornamental pine trees. Anna's dine on the nectar of exotic eucalyptus, German ivy, bottlebrush, and mallow. They have also expanded north in recent decades to the suburbs of Seattle, where winter-blooming flowers are increasingly available. The northernmost nest record for this wandering species is now from British Columbia.

It should be noted that native flowers do not bloom in expansion areas in winter. When cold snaps are severe enough to freeze the ornamentals, hummingbirds must find alternate sources of nectar and insects. Hummers of expansion areas cannot migrate to avoid fickle weather because they do not have adequate fat reserves. If subfreezing temperatures last only a day or two, hummers may be able to enter torpor to pass the worst weather. They must find feeders or fresh blossoms immediately when they awaken, however, or they will perish.

Humans who host vagrant hummingbirds have devised ingenious ways to offer nectar to their charges in areas where winter produces sustained subfreezing temperatures. During the winter of 1989–90 a rufous hummingbird stayed in Louisville, Kentucky. Its hosts placed a feeder in a box warmed by three light bulbs. Another rufous overwintering in West Virginia was kept alive by attaching a vented tin can to the bottom of the feeder and wiring a Christmas light inside to warm the contraption. Although it takes lots of effort, many winter hosts alternate two feeders—bringing cold nectar in before it freezes and replacing it with a warmer bottle.

Precautions are necessary for the more elaborate set-ups. A rescuer tried to keep an immature black-chin alive during freezing weather in Amarillo, Texas, in December 1989. The bird was coaxed into a horse barn, where its feeder was warmed by a heat lamp. Unfortunately, some of the bird's excretia hit the hot bulb, the lamp exploded, the horses bolted, and the black-chin was never seen again.

Ruby-throated hummingbirds follow traditional north-in-spring, south-in-fall migratory routes, but several western species have individual patterns. Calliope and rufous hummers, northbound from Mexico, travel through deserts of western Arizona and California in February and March. They go north along the California coast. In August they begin to return south through the Sierras and Rocky Mountains. Adult males migrate a couple of weeks before females and immature birds. Their elliptical route coincides with peak wildflower blooms, first in the desert, then in the mountains.

Costa's hummingbirds are desert dwellers, wintering where the corners of California, Arizona, and the Mexican Baja meet. Their migration is primarily east-west instead of north-south. Costa's hummers fly to Arizona beginning in late January. There, males course over desert washes performing a persistent whistle to attract mates. Females make nests decorated with silvery sage leaves in sycamore trees along riparian corridors. They are usually finished raising families by May, with most leaving the deserts for Baja country and the Pacific Coast of southern California.

Anna's hummingbirds also move along an east-west vector from southern California into central and southern Arizona in the fall. They remain during the winter, with some birds nesting. Most leave during the late spring and summer. An Anna's hummer banded at Sonoita (southeast of Tucson) was recovered six months later about 450 miles northwest in California. Anna's hummers in central and northern California migrate into the mountains and north into Oregon after their early spring breeding is completed.

Broad-billed hummers range north and west in Arizona after they nest, while post-breeding magnificent hummingbirds wander northeast. Buff-bellied hummers also scatter northward after they have finished the breeding cycle.

Just as some hummers migrate from the tropics to find food and space to nest each spring, others move to greener pastures once their young have fledged.

Discovering Hummingbird Secrets

Hummingbirds are not large enough for researchers to use tracking techniques involving radio transmitters or colored neck, wing, or tail markers. Hummingbirds do not flock, and movements of individuals are too minute to be picked up on radar, a technique successfully used to track migrant songbirds. Although information is still very sketchy, some of the best data on hummingbird movements results from banding projects.

For seventy years ornithologists have been studying all types of birds by placing lightweight numbered aluminum leg bands on individuals they briefly capture. Details about the bird's sex, age, and location are sent to the Bird Banding Laboratory at Laurel, Maryland. In 1992, computer files there contained nearly fifty million entries. When banded birds are recaptured or found dead, numbers on their leg markers are sent to the lab, adding to scientific knowledge of where the bird traveled or how long it lived.

There are 2,100 licensed bird banders in the United States, but fewer than forty have qualified as master banders of hummingbirds. The miniature size of hummingbirds—and the even tinier size of bands that fit without injuring them—requires banders to be skillful and patient. While most bird banders receive preformed bands of various sizes from the Bird Banding Lab, even the smallest sizes are too large for hummingbirds.

Years ago hummer bands were rather clumsily cut down from songbird bands. Now hummingbird banders cut and roll their own from sheets of pliable aluminum, each containing 420 sequential numbers. Black numerals are anodized to the aluminum in a six-figure code. Even for experienced hands, forming the ringlets is a tedious and time-consuming process.

Last summer I watched Sheri Williamson, an avian specialist and former zookeeper, prepare for hummingbird banding at Ramsey Canyon, Arizona. At this Nature Conservancy preserve, hummers are captured one morning and evening each week from late spring through fall. Guests staying on the sanctuary grounds are welcome to observe.

Using special shears, Sheri snipped the aluminum sheet along hatch marks. For hummers the size of black-chins and rufous, bands of 0.08-inch outer diameter are required. Bands 0.1 inch in diameter fit magnificent and blue-throated hummingbirds. Tolerances in the length and width of the bands must be exact within thousands of an inch. Hummer banders use precision-engineered tools, of which no more than three dozen sets exist, to shape metal strips.

When laid on top of a penny, the smaller band occupies the area between Lincoln's nose and chin. Sheri smoothed the edges of her bands with carborundum paper so metal burrs would not stab delicate legs.

Ruth Russell, who has banded hummingbirds since 1985, works on one of the most ambitious banding programs in Arizona. She is assisted by her husband Steve, a professor of ornithology at University of Arizona, and several volunteers. For

The miniature size of hummingbirds does not allow researchers to track them with radio telemetry. A few skillful master banders across the country do capture and band hummingbirds to gather information about their life histories. (Photo copyright by Connie Toops)

Bands sized for magnificent and blue-throated hummers are 0.1 inch in diameter. Those worn by hummingbirds the size of ruby-throats or rufous are even tinier. (Photo copyright by Connie Toops)

the past five years, Ruth has spent one day a week from late spring through fall netting humming-birds near Sonoita. During peak migration periods she bands there more frequently. On Memorial and Labor Day weekends she coordinates with other researchers by banding at a site near the Canelo Hills Nature Conservancy preserve. She invited me to join her on Labor Day in 1991.

I was unsure I had taken the proper fork of the rural road until I saw a cluster of cars near a ranch house and two women running across the lawn waving their hands in the air. They were not flagging me down; they were hummingbird wranglers.

The Russells trap birds with a rectangular enclosure of mesh netting suspended on upright poles. They place two hummingbird feeders inside near the opening of the net. Since all other feeders at the site are removed, hungry hummers enter the corral for a drink. Wranglers wait until the birds sip nectar, then run toward the opening, forcing the hummers farther into the trap. When birds are ensnared in the fine mesh, wranglers swiftly pluck them out and ready them for banding.

On the screened porch Ruth, Steve, and their assistants employed a well-organized data collection process that took less than two minutes per bird. Netted hummers were placed in mesh bags and brought to the porch in order of capture.

I gazed over Ruth's shoulder as she accepted a bagged hummer from the wranglers. "Adult male black-chin," she said as Steve typed codes into the computer. Ruth gently placed a band around its tiny leg and closed it with special pliers that do not pinch. "T45324," she read from the band. With the bird in her left hand, she measured its wing, tail, and beak, telling Steve each length. Looking through her magnifier at the bird's beak, Ruth said, "Smooth." (Adult hummers do not have wrinkled grooves on their bills as the bills of immature birds do.)

Continuing the examination, Ruth enumer-

A black-chinned hummingbird in southeastern Arizona receives a numbered leg band. It may someday be recovered to help researchers learn where these birds migrate and how long they live. (Photo copyright by Connie Toops)

ated, "Ninety-nine percent, one-ventral, one-dorsal, light, fresh, fresh." This told Steve the bird's gorget was fully feathered, what state of feather molt the bird was in, and how worn the feathers were. Ruth draped nylon netting around the hummer to temporarily immobilize it and laid it on a scale. "Three point two grams," Steve read.

Finally, two volunteers served "lunch." They retrieved each bird from the scale and encouraged it to drink from a feeder. After an energy-restoring gulp, each hummer was released.

Later on this Labor Day weekend, I drove to Madera Canyon, where David and Linda Ferry were also banding hummingbirds. The Ferrys, both medical doctors, volunteered to study hummingbirds at Ramsey Canyon, Arizona, in 1986. A year later they added a five-year study of Madera Canyon's hummers as a comparison. Each Memorial and Labor Day weekend since, plus two weeks in July at the peak of the reproductive season, they have netted and banded hummingbirds.

The Ferrys trap birds with enclosures similar to the Russells' traps. When I met them at Santa Rita Lodge, they had four nets open and a dozen experienced volunteers as wranglers. Linda and David took turns examining the birds at a work station on the porch of their cabin.

Volunteers retrieved birds from the traps and transported them to Linda or David in net-sided boxes. I watched as Linda, a petite, soft-spoken woman, placed a wisp of nylon mesh around each bird and weighed it. She placed a band on the left leg of a new individual or read the number on the band of a recapture. She also measured bill, wing, and tail lengths and noted feather condition. In moments, each bird was free again.

Both the Russells and the Ferrys invited me to join their wranglers. The first hummer I nabbed was a young black-chin. Having banded warblers previously, I thought I was prepared. I gingerly placed my hand around the black-chin's buzzing wings to quiet it while I unlaced the fine netting

Left: Hummingbirds are trapped in enclosures of fine mesh netting. The entire process—from capture, through banding and weighing, to release—takes about five minutes. (Photo copyright by Connie Toops)

Above: Magnificent hummingbirds sometimes sit quietly for a few moments after banding, unaware they are free to fly. (Photo copyright by Connie Toops)

Bands, such as the one worn by this broad-billed hummingbird, are painstakingly fashioned from sheets of pliable aluminum supplied by the Bird Banding Laboratory at Laurel, Maryland. (Photo copyright by Charles W. Melton)

from its head and neck. "*Tchew, tchew, chupa-tchew,*" it scolded. I loosened my grip. Afraid that I had squeezed the hummer too hard, I grasped it again, even more delicately. "*Tchew, tchew,*" it cried, fluttering from my fingers but still confined in the net. Another wrangler offered assistance.

"These black-chins complain a lot," the wrangler said as she cupped it in one hand and pulled the loops of net over its head with the other. "They're tougher than they look, though. You just get a good grip and back them out of the net."

I rescued the next bird without ado. Black-chins were always feisty, greeting capture with a tirade of squeaky complaints. They were most numerous at the Russells' site. In Madera Canyon I plucked black-chins, colorful broad-bills, and a few Anna's from the net. Then a magnificent hummer landed.

I sprinted toward the trap, waving my arms. The magnificent flopped into loose folds of the mist net. Within seconds, it was in my hand. I

marveled at the shimmering purple of its cap and green of its gorget. This bird was twice the size of previous captures. Even so, I could barely feel any weight as I held it.

Linda measured the magnificent hummer, then put the banded bird in my hand for release. It sat there on my outstretched fingers for a few moments, unaware it could go. I witnessed this same calm demeanor in several magnificent hummingbirds Sheri Williamson banded at Ramsey Canyon. Finally it realized it was free. In less time than it takes to recount, it rose from my hand and disappeared into the juniper forest, leaving a warm glow where it had rested.

Banders such as the Ferrys and Sheri Williamson, who work in areas open to the public, are sometimes asked why it is necessary to capture birds and subject them to the indignities of being

75

weighed and measured. David Ferry explained, "Unless birds can be identified as individuals, no information regarding local movements, length of time spent in the breeding territory, the numbers returning to the same breeding site year after year, or life expectancy can be obtained."

Although hummer banding is relatively new compared to general bird banding, researchers discover more about these intriguing creatures with each passing year. As the Ferrys' project ended in 1991, they had netted about 5,500 hummingbirds in southeastern Arizona. Highlights included the first United States captures of berylline, violet-crowned, and white-eared hummers and banding more broad-billed, blue-throated, and magnificent hummers than the total of all previous studies. They have statistically documented the recent influx of Anna's hummingbirds to Arizona, especially at Ramsey Canyon, and have identified in-hand more Allen's hummers than were previously thought to migrate through the region.

Among their interesting recaptures were a violet-crowned that returned to Ramsey Canyon eleven months after it was banded there, and a young blue-throated hummingbird banded at Ramsey and then caught at Madera only twenty-six hours later. A female berylline traveled back and forth between Ramsey and Madera canyons twice in the summer of 1987.

Bill Calder is a pioneering hummingbird bander. His experience spans twenty years of study, and he has helped many other banders start their projects. Calder captures nesting broad-tails and migrant rufous hummers each summer in Gothic, Colorado. He travels to Mexico to search for them in the winter. To date he has not located a hummer on both its summering and wintering sites, but he has nabbed at Gothic a rufous banded in Montana and another banded in southwestern New Mexico. Many of his recapture dates are within a week of original bandings at Gothic in a previous year.

This site fidelity is common for hummingbirds. Most of Ruth Russell's recaptures are also at the same place, near the date they were banded in a previous year. With the help of assistant Susan Wethington, Russell's studies have revealed that the same hummingbird will use feeders within two miles of each other. Feeders five miles apart are used by separate hummer populations.

Marguerite Baumgartner, who banded for a dozen years in northeastern Oklahoma, identified an amazing 452 individual ruby-throated hummers using her yard in 1987. Nancy Newfield has substantiated similar abundances in small yards. At one residence where hosts thought they had about twenty birds, Newfield banded eighty-seven. A safe rule of thumb, Nancy says, is to multiply the number of birds you can count at one time by four.

Newfield also made some interesting discoveries after the dry midwestern summer of 1983. Since hummingbirds rely on feeders more during dry periods, Newfield banded about twice the normal number of southbound ruby-throats that year. She noted, however, that these birds were not as fat as when midwestern wildflowers were abundant. She recorded a lower than normal percentage of adult ruby-throats returning the following spring.

The most intriguing secrets remaining for hummingbird banders to discover are migration routes between summering and wintering areas. As more banders mark more birds, bits of information will begin to fall into place. Until then, a few aspects of hummingbird migration remain a mystery.

Recaptures of banded hummingbirds help determine their life spans. The oldest wild black-chinned hummingbirds recovered thus far were about five years old. (Photo copyright by Connie Toops)

Meccas of Hummingbirding

The terrain of southeast Arizona consists of scattered peaks rather than one mountain range. From Tucson south to the Mexican border and east to New Mexico, ten massive "sky islands" rise from surrounding desert plateaus. With summits over nine thousand feet, mountain tops are cool while the desert below bakes in sweltering sun.

Above the cacti and gray-green creosote bushes of the lowlands are transition forests of juniper and oak. They lead to verdant pines, firs, and spruces on the highest peaks. Water from rain and snow courses down rugged canyons lined with Arizona sycamore and walnut, bigtooth maple, ash trees, and a profusion of wildflowers. The flora and fauna of the sky islands are similar to northern Mexico.

For birders, the area offers specialties including elegant trogons, sulphur-bellied flycatchers, and red-faced warblers. In midsummer seven primarily Mexican species of hummingbirds inhabit the sky islands. They include magnificent, blue-throated, and broad-billed hummingbirds—all relatively easy to find in proper habitats—and more elusive white-eared, violet-crowned, Lucifer, and berylline hummers. The plain-capped starthroat visits on rare occasions.

Northward hummingbird migration reaches a crescendo in early May. Local birds nest in June and July. A summer rainy period occurs in late July and August, bringing new wildflowers and second nestings for many species. Returning migrants plus fledging juveniles make late summer a prime time for hummer abundance and diversity.

RAMSEY CANYON, ARIZONA

I arrived at the Tucson airport on an August afternoon that would have made a tour of a blast furnace feel refreshing. My goal was the Huachuca Mountains, about ninety miles distant on the southeast horizon. As the afternoon waned, I drove across desolate washes edged by squatty, thorn-covered trees. Between them were grasslands studded with clumps of sharp agaves. It was dusk when I reached Ramsey Canyon. Cool, moist air rushed down the streambed at this 5,400-foot elevation, where temperatures are often fifteen degrees lower than in Tucson.

My stay began at Ramsey Canyon Inn, a bed and breakfast adjacent to the Mile-Hi Nature Conservancy Preserve. At first light the next morning I saw an array of feeders surrounding the house and dangling from venerable oaks on the lawn. Bright red flowers, in beds and clay pots, decorated the yard.

I settled into a porch chair as several black-chinned hummingbirds swarmed to the feeders. When males turned at just the right angle, their gorgets glowed with rich purple iridescence. Female and immature black-chins uttered soft *tchew* calls and bobbed their tails as they drank. Whooshing wingbeats announced the arrival of a blue-throat, a hummingbird the size of a sparrow! Compared to the soft whine of the black-chin's wings, it sounded like a thundering jet.

I wandered behind the inn, where an adult male magnificent hummingbird zoomed to a feeder. As it banked, sun glistened on its purple cap. On its vel-

In the summer, seven species of hummingbirds from Mexico may be observed in the mountains of southeastern Arizona. (Photo copyright by Connie Toops)

vety black breast was a green gorget glowing like a traffic light on a rainy night. Female or immature magnificent and blue-throated hummers can be confused, but an excellent identifier is the amount of white on the tail. Blue-throats have broad tails with large white patches on the outer corners. Their call is a loud, explosive *seep* rather than the thin *chip* of a magnificent.

Two birders from Washington, Mike and Paul, emerged from the inn. They hoped to encounter eared trogons reported farther up Ramsey Canyon and white-eared hummingbirds occasionally seen at feeders here. More abundant in highlands of northern and central Mexico, white-eared hummers have summered in southern Arizona, New Mexico, and Texas. They are most likely to be seen near streams in pine-oak woodlands.

We watched magnificent, blue-throated, black-chinned, and Anna's hummers dine at the feeder. Then I saw a small bird land high in the tree. I raised my binoculars and recognized it as a long-awaited jewel from the "southwestern specialties" page of my bird guide. It was a male white-eared hummingbird in all its finery—jaunty red beak, crisp white eye line, and when it looked directly at me, shimmering violet feathers on the crown and chin. "There it is!" I gasped.

We studied the stocky white-ear as it fed and returned to the branch twice. In less than two minutes, it vanished. We trouped jubilantly into the kitchen for breakfast.

Innkeeper Shirlene Milligan's dining table overlooks the back yard. Attached to the window is a small vial of sugar water. None of us had taken more than ten bites of our omelets when all forks froze in mid-air. A white-eared hummer landed on the window feeder, only inches from us. Mike glimpsed a band on its leg. It was a fledgling male from a nest at Mile-Hi Preserve. One brief sip and it departed.

Later I walked up the road to Mile-Hi, site of the fourth known berylline hummingbird nest in the United States. Interest in it was so intense that birders were arriving from all over the country. Preserve staffers mounted a spotting scope at the edge of the parking lot, focused on a tiny cup adorned with bits of lichen. I peered through the scope, noting the green head of the incubating bird. In my first six hours of hummingbirding at Ramsey Canyon, I encountered five life species!

Shirlene served breakfast at 8:00 A.M. Conversation at the table the next morning centered on birds, since most guests were birders. But at 8:12, talk ceased and all eyes focused on the window feeder. The young white-eared hummingbird was back, within a minute or two of the time we saw it yesterday. We agreed Shirlene's inn would always be full if she could guarantee a white-eared hummer with breakfast.

I strolled to Mile-Hi Preserve for another look at the berylline hummingbird. The spotting scope was still trained on the nest, but today the lichen-covered abode was empty.

Naturalist Jack Whetstone explained, "The female berylline has been sitting on this nest twenty-eight days. Normally eggs would hatch in a little over two weeks. The most recent sighting of a male berylline was three months ago. We speculate the female mated with another species and her eggs were infertile.

"Mating between species takes place in hummingbirds more often than you might suspect," Jack continued. "In rare instances it produces hybrid offspring. Usually, the eggs fail to hatch."

Birders gathered in front of the visitor center, listening to Jack. He gestured toward a red quart feeder in an apple tree the female berylline used. "This morning she has appeared about every fifteen minutes," Whetstone said.

Half a dozen of us lined up ten yards away, raising our binoculars expectantly whenever a bird buzzed to the feeder. In a few minutes the berylline arrived. She was larger than an Anna's hummingbird, and her most striking features were a green head and throat. I was close enough now to distinguish red at the base of her lower mandible. She sipped nectar for a moment, then departed in a flash of chestnut, the color of the undersides of her wings.

When male magnificent hummingbirds turn at just the right angle, their caps glow rich purple and their gorgets are neon green. (Photo copyright by William B. Folsom)

The Chiricahua Mountains form the largest of southeastern Arizona's sky islands. They host more types of breeding birds than any other region in the United States. (Photo copyright by Connie Toops)

From the inn, I moved to Sycamore Cabin at Mile-Hi Preserve. The main portion of the sanctuary, a 280-acre tract within Coronado National Forest, was donated by Dr. Nelson Bledsoe in 1974. A year later, six streamfront vacation cabins on an adjoining twenty-acre plot where the visitor center is now located were purchased by the Nature Conservancy.

The preserve's terrain rises from 5,500 to 6,400 feet. It varies from a lush riparian corridor to sparse cacti, yuccas, and dense pine-fir forests. This diversity has enticed, at one time or another, fifteen species of hummingbirds. In late July 1989, eleven hummer species were observed on one memorable day.

Despite the possibility of seeing trogons, eagles, painted redstarts, or mountain lions, the main drawing card for the sanctuary is its array and abundance of hummingbirds. Inside the front gate, under a shade tree, are several benches facing a row of hummingbird feeders. They dangle over a bed of Lemmon's sage, betony, and other native red flowers.

From 8:00 A.M. to 5:00 P.M., spring through fall, hummer-watchers flock to the benches. Newcomers, armed with binoculars and guide books, patiently sort characteristics to identify each species. Return visitors are less concerned about names. They may sit quietly for half an hour to half a day, entranced by the aerial antics and dazzling colors of these flying jewels.

Each cabin has its own feeders, native plants, and lawn chairs. Guests often have cameras or binoculars in hand, enjoying a private audience with the birds. Some can observe female blue-throated hummingbirds feeding babies, since many of the cabins have blue-throat nests under the eaves.

Blue-throated hummingbirds inhabit mountain canyons with flowing water. They are widespread in the Sierra Madre of western and central Mexico. They frequent the maple-sycamore streamside

A few white-eared hummingbirds from Mexico spend the summer along wooded streams in southern Arizona, New Mexico, and western Texas. (Photo copyright by Clayton A. Fogle)

Cave Creek, a lush riparian corridor in the Chiricahua Mountains, provides a sharp contrast to the surrounding Arizona desert. Blue-throated hummingbirds nest here in summer. Watchful birders may also spot elegant trogons and painted redstarts. (Photo copyright by Connie Toops)

Costa's hummingbirds are numerous in southern Arizona in the spring. Males are distinguished by long purple gorget feathers that extend beyond each side of the neck. (Photo copyright by Connie Toops)

habitat of Ramsey and Cave Creek canyons each summer.

Near the end of my stay, I witnessed the cooperation between business and scientific communities fostered by the Nature Conservancy. At 6:30 A.M. Bob Jones, a lineman for the Sulfur Springs Valley Electric Coop, came rumbling up the narrow dirt road in his big truck. A Conservancy official had alerted a director at the power company of the importance of the rare berylline nest. Jones was sent to help collect it.

Preserve manager Tom Wood pointed out the tiny nest to Jones, who carefully maneuvered his truck into position. The lineman raised the bucket high amid the sycamore branches, where he photographed the nest. Then, under Wood's direction, he clipped the branch to which it was attached and brought it the ground. After the female abandoned the nest, her eggs had been claimed by a predator, but the delicate cup was in perfect condition. It will take a place of honor in a study collection as

only the second berylline hummingbird nest collected in the United States.

CAVE CREEK CANYON, ARIZONA

From Ramsey Canyon I drove east to Cave Creek Canyon, near the Arizona–New Mexico state line. From a distance, the mouth of the canyon is marked by imposing walls of rhyolite. Beneath these pastel cliffs lies the village of Portal. Multiple hummingbird feeders hang outside the combined general store–restaurant–bed and breakfast. Many birders pause here to check a logbook of rare bird observations left by previous visitors.

Vegetation along the road to Portal is sparse, but beyond the village a lush creekside corridor of white-trunked sycamores winds upward to the evergreen oaks and pines of higher elevations. Not too far beyond Portal, a gravel road branches left. I followed it a third of a mile to a driveway marked "Birders Welcome." I parked outside the gate, walked quietly around the side of the stucco home,

Sightings of violet-crowned hummingbirds have become more common in southeastern Arizona in the past three decades. (Photo copyright by Charles W. Melton)

and took a seat.

I was a guest of Drs. Walter and Sally Spofford, known internationally for the birds and birders that visit their yard. Sally retired from the ornithology staff at Cornell in 1969. She and Walter purchased their delightful homesite along Cave Creek two decades ago and recently donated an easement to the Arizona Nature Conservancy for The Portal Preserve.

Chairs and benches to seat twenty people were arranged in a semicircle. They faced more than a dozen hummingbird feeders of all shapes and sizes hanging under the eaves and on a wire stretched across the yard. Seed feeders for finches, suet-cornmeal-peanut butter cakes for woodpeckers, brush piles to shelter sparrows and quail, and pans of water were liberally distributed beneath the trees.

On a stump I found a guest register and brief note of welcome. The Spoffords apologized for not greeting each of us, but with four thousand visitors a year, personal hellos are infeasible. Birders have found their way to the yard from all fifty states and twenty-nine nations. Through the years, 210 species of birds have been spotted here, including twelve species of hummingbirds.

Feeders need to be refilled from one to four times daily, depending upon the season. At peak periods, Sally serves two or three gallons of sugar water per day to hummers, finches, orioles, and even nectar-eating bats. When the Spoffords are away, they employ house sitters to care for the birds.

A male Lucifer hummingbird had been seen at the Spoffords' feeders recently, so I settled onto a bench to wait for it. House finches and a black-headed grosbeak dined on sunflower seeds and milo. Black-chinned hummingbirds darted to the hanging feeders. An Anna's joined them, then a male blue-throat swooped to the porch. A troupe of Mexican jays stared down at me from the spreading branches of a nearby oak.

Hummingbirds and birders flock to the feeders of Wally and Marion Paton in Patagonia, Arizona. (Photo copyright by Connie Toops)

Lucifer hummingbirds are uncommon summer visitors to southwestern Texas and southeastern Arizona. Their bills are more down-curved than those of other North American hummers. (Photo copyright by Greg W. Lasley)

Two more birders slipped into the yard. "Have you seen the Lucifer?" the woman whispered. "Not yet," I replied.

We watched the jays, finches, grosbeaks, and black-chins in silence. About forty minutes after my arrival, a wisp of a hummer landed on the wire from which the feeders were suspended. A look through my binoculars confirmed a slightly downcurved bill and a shiny purple gorget that trailed into individual feather tabs on either side of the bird's neck. It fluttered to the feeder, gulped a few times, and departed.

I stayed a bit longer. Two boldly marked acorn woodpeckers relayed up and down an old agave stalk, searching for insects and suet. A third hung on the fence near a hummingbird feeder and comically stretched its neck to lap sugary liquid from the port. Then a roadrunner stepped out of the brush and walked directly to the patio door, where it tapped on the glass. To my surprise, a young woman opened the door and tossed two chunks of beef to the bird. The house sitter saw me and seemed obliged to explain.

"The roadrunner is a regular visitor," she stated. "Unfortunately, when we're not around, it jumps up and grabs birds. Sally has a picture of it leaping off the roof and catching a black-chinned hummingbird at one of the feeders." Tossing more meat, she added, "We keep it full so it won't bother the other birds."

The woman returned to the house and the roadrunner walked to the middle of the yard. For several minutes it eyed the comings and goings of unconcerned hummingbirds. If one perched for more than a few seconds, the roadrunner cocked its head and stared. I wondered when the Lucifer would return. I also wondered whether the roadrunner made any distinction between common and gourmet hummers. I was relieved when it stalked back into the brush.

About forty minutes after its first appearance, the Lucifer again landed on the wire and scanned the yard. This time the dominant blue-throat sliced through the air and buzzed the Lucifer. Flaring its forked tail and wheeling quickly, the Lucifer disappeared over the shrubs along the west side of the yard.

I was to stay four miles farther up Cave Creek at Southwest Research Station, a scientific outpost established in 1955 by the American Museum of Natural History. Several cabins, an office, and a lab are nestled peacefully below the weathered pink bluffs. In the communal dining hall, I met graduate students and university professors studying such diverse subjects as geology, butterflies, lizards, and mushrooms.

The Chiricahua Mountains, of which Cave Creek is a major drainage, form Arizona's largest sky island. For plants and animals, they are a transitional bridge between Mexico's Sierra Madre Occidental and the Rockies. They also lie between the Sonoran and Chiricahuan deserts. Naturalists have recognized the area's biodiversity for decades. A staggering 25,000 species of insects dwell here, as do nearly 1,400 kinds of flowering plants. More types of birds breed in the Chiricahuas than in any other region of the United States.

When a staff member learned I was interested in hummers, she phoned a friend who had a violet-crowned hummingbird coming to his yard. He was leaving early the next morning to lead a birding tour but welcomed me to come look in exchange for putting fresh sugar water in the feeders.

The residence was near the mouth of the canyon. I found four feeders suspended above a fence that separated the yard from a brushy field. Two others dangled from branches of shade trees. The first to return after I filled the feeders was a black-chinned hummingbird.

A sharp *seep* sounded in the oak tree over my head and a sleek form burst into flight, zipping through fifty feet of open air to the feeder and diving on the black-chin. The attacker was a male blue-throated hummingbird. It returned to a perch ten feet above me, but had I not watched the bird land, I probably would not have noticed it.

Blue-throated hummingbirds are drab compared to other Arizona hummers. Although males have a shiny blue gorget when seen at close range, the overall impression of sitting birds is one of dull green backs, grayish underparts, and dark tails.

Since they spend much of the day perched in the shady canopy of riparian woodlands, they do not need to advertise their presence with color as hummers of open areas do.

If a smaller hummer approached one of the feeders, the blue-throat raced off to challenge it. It perched on low branches in each of three large trees in the yard. Every spot allowed the blue-throat to sit in obscurity, yet afforded a full view of all feeders. Not only did the domineering blue-throat swoop on and chase away hummingbirds. It faced off with a house finch, razzed a red-tailed hawk that paused on a snag at the edge of the field, and regularly challenged bees and wasps hovering around the feeders. The blue-throat's sleek wings and its swift flight reminded me of a fighter plane.

The blue-throat ruled the yard, but smaller hummingbirds obtained a share of nectar by darting to the farthest feeder. If they sat quietly, they usually went undetected. When they hovered or came closer, the blue-throat attacked.

I glimpsed a hummer with an unusually white breast in the trees at the back of the field. It was the elusive violet-crowned. But the blue-throat also saw it and chased it away. A few moments later another small hummer relayed from the back of the field. Wary of the blue-throat, it sneaked in for a quick sip. It had a slightly curved bill and distinctly forked tail. It was a male Lucifer!

Shortly the female Lucifer made a direct approach from the field, pausing on a low bush just behind the fence. I saw her buffy breast, but so did the blue-throat. He chased her back to the trees.

I left the yard as a rainshower arrived but passed the driveway later in the afternoon. I decided to try again for a good look at the violet-crowned.

As the sun circled west, so had the blue-throat, keeping the light behind it. An immature male blue-throat arrived and the dominant bird dove on it, flaring its tail to show wide white spots at the outer edges. The intruder cut right, heading straight for the house. Skyhawk followed. My lawn chair was near the wall, and as the birds turned, there was no way for them to go but straight toward me. I braced for impact, but amazingly both went under my seat and zipped out the

other side. The intruder disappeared. The warlord paused at a feeder, then returned to its perch with a loud *seep*.

Minutes later the young blue-throat tried again. The dominant bird often dove on intruders, forcing them down. This time he caught the interloper. For a few seconds they tussled on the ground in a fluttering ball of feathers. Disheveled but defiant, the ruling blue-throat rose to its perch and uttered a ticking, wheezy tirade. I could imitate it by placing the tip of my tongue behind my front teeth and drawing in short breaths of air. After vocalizing, the blue-throat spent ten minutes preening.

In the meantime, a huge thunderstorm billowed over the rocky slopes of Cave Creek Canyon. As is typical of the summer monsoon season, puffy white clouds boiled into ominous gray masses. Distant thunder rumbled between the rock walls. The breeze, which had blown all day, stopped.

A male rufous hummingbird flew to a feeder. It was probably a passing migrant, as I had not seen the distinctive russet bird previously. Had the blue-throat met a match in this feisty hummer? It allowed the rufous to drink its fill.

As the storm rolled nearer, I doubted that theory. The sky darkened and thunder arrived with a sharp crack. All kinds of hummers were coming into the feeders in droves now. Obviously, they wanted to eat before the sheets of cold rain arrived. The blue-throat chased a few birds but allowed most to drink heartily. The male Lucifer came twice within a few minutes.

Lightning flashed. I counted seconds until the sound of the thunder reached me. The storm was only a mile away. Every feeder had multiple hummingbirds on it now. I smelled rain.

Suddenly, there it was—a vibrant male violet-crowned hummingbird on a mesquite tree behind the fence. It turned its head watchfully from side to side. With each movement, the cap radiated rich purple. To the feeder it fluttered, taking long, sweet drinks. Through binoculars I studied the clear white breast, the olive-brown back, the brilliant cap, the red rapier beak tipped in black. I savored every second the bird remained at the

Magnificent hummingbirds frequent feeders at Arizona birding meccas including the Mile-Hi Preserve at Ramsey Canyon and Santa Rita Lodge in Madera Canyon. (Photo copyright by Sid Rucker)

Broad-billed hummingbirds abound at Madera Canyon. Their wide red bills are tipped black. (Photo copyright by Sid Rucker)

feeder. Then it flew, and with its departure came the rain.

PATAGONIA, ARIZONA

Seeing the violet-crowned hummingbird in Cave Creek Canyon was an unexpected surprise. Others had recommended Wally Paton's yard in Patagonia as *the* place to see a violet-crowned, the rarest hummingbird regularly nesting in southeastern Arizona.

Patagonia lies twenty miles northeast of Nogales, Mexico, on Arizona Highway 82. To reach Paton's, I turned onto Pennsylvania Street, then crossed Sonoita Creek at the low water bridge. On the left I noted a gate with the sign, "Hummingbirders Welcome." A path led across the lawn. Behind the house, several chairs faced a bank of feeders hanging outside the kitchen window.

Each feeder was abuzz with multiple hummers scrapping over perches. Broad-billed hummingbirds were abundant. A birder gathering his camera gear told me, "The violet-crowned was just here. It should be back within fifteen minutes."

True to his word, a female violet-crowned soon landed on a feeder twenty feet in front of me. Its head was drabber than a mature male's, but the white breast and black-tipped red beak were distinctive. Later, an immature male came for a drink.

About that time a slim gentleman emerged from the back porch with fresh feeders in hand. I introduced myself to Wally Paton and then met his wife Marion. I learned they feed forty-five gallons of sugar water a month in peak seasons.

The Patons' large yard stretches to the banks of Sonoita Creek. This riparian habitat, and the Patons' proximity to the Nature Conservancy preserve upstream, is responsible for the half-dozen violet-crowned hummers they see at their feeders. In west-central Mexico, where the birds are more abundant, they associate with streamside cottonwood, willow, and sycamore trees.

When Herbert Brandt published his treatise on Arizona birds, he knew of only two violet-crowned sightings in the state prior to 1951. In 1957 ornithologists spotted six of the unusual hummers at Guadalupe Canyon on the Arizona–New Mexico border. Two years later, they discovered four violet-crowned nests—all in sycamore trees. Since then violet-crowns have occasionally been sighted at Cave Creek, Ramsey, and Madera canyons and along Sonoita Creek.

Some Mexican hummers are so scarce in the United States that when news of a sighting spreads on the birding grapevine, hosts have more onlookers than they bargain for. When a plain-capped starthroat arrived at another residence in Patagonia in 1978, so many birders flocked to see it, the sheriff had to clear cars off the street.

Luckily Wally and Marion enjoy their role as hummer hosts. "We meet some interesting people," Wally said in his crisp Boston accent. Marion recalled when word about their violet-crowned hummers first went out, she noticed people climbing on the fence and in the trees behind the yard to catch a glimpse of them.

In 1990 the Patons began inviting birders into the yard. A year later they turned a former garden into lawn space, planting Lemmon's sage and trumpet vine to attract more birds. Wally believes they host up to two thousand birders a year, and

his guest register shows visitors not only from the United States and Canada but from Australia, New Zealand, Great Britain, Belgium, and Africa.

Bird hosts including the Patons and Spoffords have a few simple requests. It is not necessary to call first, but do come at reasonable times. Hours between 8:00 A.M. and 5:00 P.M. are acceptable to most. Park so you do not block traffic or private driveways. Sign the guest register, but do not remain too long or expect personal greetings. If you photograph the bird, send a print as a thank you. Although they may not ask, leave a contribution to help defray costs of maintaining their feeders.

MADERA CANYON, ARIZONA

From Patagonia I continued my hummingbird odyssey by driving to Madera Canyon. My destination was Santa Rita Lodge, a homey cluster of cabins operated by David and Lyle Collister within Coronado National Forest. The resort is nestled among live oaks and alligator junipers at about five thousand feet. It overlooks Madera Creek. Among the jumbled boulders of its banks are red chuparosa, paintbrush, and verbena flowers. The attraction of flowers and nest sites along the stream and the "couple of hundred pounds of sugar per month" Lyle reports she offers in feeders around the lodge, result in an abundance of hummingbirds.

Each room or cabin has a hummingbird feeder of its own. Fascinated, I did not bother to unpack but went straight to the window and watched a procession of broad-billed, black-chinned, rufous, magnificent, and Anna's hummingbirds. The streambank drops away quickly behind the lodge, so from my window I saw hummers flit into nearby treetops to rest and preen between meals.

Less than an hour's drive southeast of Tucson, Madera Canyon is the most-visited of Arizona's birding meccas. One of the special activities that drew me to Santa Rita Lodge at this particular time was a hummingbird workshop led by master birder Kenn Kaufman and his wife Lynn. Throughout the spring and summer the Collisters arrange weekend workshops for guests. Along with fourteen others, I met Kenn and Lynn early Sunday

morning. After a bit of general bird-watching, we sat in a semicircle around the feeders in front of the cabins. As hummers flitted in and out of view, Kenn offered tips on identifying them and elaborated on their natural history.

Among the first to frequent the feeders was a male broad-billed hummingbird. Broad-bills are common at Madera, but unusual at Ramsey Canyon, only forty miles distant. Broad-bills are slightly larger than ruby-throated or black-chinned hummingbirds. The males are plumed in shimmering green with midnight blue throats and tails. Females have green backs, drab underparts, and a small white mark behind the eye. Both sport unmistakably wide blood-red beaks tipped in black. Their call is a dry *ick-ick*. Broad-bills arrive here in early spring to nest but most return to northern and central Mexico for the winter.

Kenn helped us distinguish between the similar calls of black-chinned and rufous hummingbirds, the black-chinned's being lower in pitch. The black-chin also has a proportionally longer bill than either Anna's or rufous hummingbirds.

Rufous hummingbirds are common late summer migrants in the mountains of southern Arizona. A few migrant Allen's hummers are here at the same time. The mature male rufous has distinctive russet plumage and a peculiar in-flight buzz. Female or immature Allen's and rufous hummers cannot safely be told apart unless they are in hand. Although about 80 percent of the birds in question here are rufous hummingbirds, Kenn explained careful birders identify juveniles and females on the wing only by genus, calling them all *Selasphorus*.

We listened to Kenn but kept our eyes and binoculars trained on the feeders. Now and then an involuntary "ooh" or "ah" would rise from the crowd as a male magnificent hummer landed in light that illuminated its purple crown and neon green gorget or a male Anna's flew in with head and chin glistening rose-purple.

Kenn ended the session by reminding us that southern Arizona's variety of hummingbirds is terrain-related. The desert attracts black-chinned and Costa's hummers, broad-bills use scrub forests of middle elevations, and sky island coniferous forests are frequented by broad-tailed, blue-throated, and magnificent hummers. He also explained violet-crowned and Lucifer hummers have become more common in southeast Arizona in the past half-century. Their advance northward from Mexico has come in short flights from one suitable canyon or sky island to the next.

DESERT MUSEUM, ARIZONA

My circular route of hummingbird hot spots brought me back to Tucson, to the Arizona–Sonoran Desert Museum. Its extensive open-air collection includes many plants and animals native to the Southwest. I headed straight for the hummingbird aviary, a large screened room filled with flowers, shrubs, and numerous hummers zooming from branch to branch.

Hidden among the vegetation are three dozen small feeders filled with a commercial nectar-protein mixture. Fruit flies are released in the aviary throughout the day to supplement diets. The enclosure provides an opportunity to study hummers at close range. I watched as a male Costa's in beautiful breeding plumage defended a territory in one corner of the aviary. A young black-chin laid similar claim to the airspace surrounding its ocotillo branch. Some of the broad-billed and Costa's hummers feel so at home here they have nested and raised young.

An outdoor demonstration garden near the aviary features plants that attract hummingbirds and butterflies while requiring little water from this arid landscape. Wild hummers buzzed back and forth there.

When I first arrived, the desert seemed alien to me—barren ground, thorny vegetation, dry heat. But as I toured the sky islands in search of hummingbirds, I began to see the fragile, balanced nature of this environment. These plants and animals are resourceful survivors in a demanding habitat. I will long remember the overt beauty and the innate complexities I discovered here.

Blue-throated hummingbirds defend territories from perches around the perimeter. They will chase intruding birds as large as hawks. (Photo copyright by Sid Rucker)

The Hummer Celebration

"Third Annual Hummer/Bird Celebration," read an intriguing notice in a birding magazine. Curiosity aroused, I phoned the number listed for the Rockport Chamber of Commerce and my inquiry was directed to Betty Baker, chairperson of the 1991 gathering.

In an enthusiastic drawl, Betty explained that Rockport and Fulton, Texas, lie along the migratory flyway for ruby-throated hummingbirds. During the September migration, fifty to two hundred of the little travelers may dart about at residential feeders or in lush stands of Turk's cap, local red wildflowers. Hummingbird experts speak at the celebration's workshops and area residents open their yards so visitors can watch the migrants. Betty invited me to join them.

Massed ruby-throats were first observed by "Texas Bird Lady" Connie Hagar half a century ago. On the coastal bend, where migratory flyways from the midwestern and west-central United States converge, there is a narrow strip of live oak woodland that separates dry interior brushlands from waters of the Gulf. This corridor of forest cover acts as a funnel, directing broad-winged hawks, colorful warblers, and hummingbirds north in the spring and south each autumn.

In 1938 Mrs. Hagar was amazed to see swarms of ruby-throated hummingbirds. "One can sit all day . . . and watch the birds flying into and out of blossoms, perching on small twigs, and catching minute insects by running out their long tubular tongues," she wrote.

Mrs. Hagar learned southbound birds arrived at the upper end of Live Oak Peninsula, which divides Copano and Aransas bays. She watched the hummers feed and move southeast to Mustang and Padre islands, offshore from Corpus Christi.

Although ruby-throats are the most common hummingbirds along the Texas coast, the area lies at the edge of the black-chin's breeding range and rarely hosts rufous hummers. Connie Hagar's careful observations revealed a few visiting Anna's, broad-tailed, and Costa's hummingbirds. The mass hummingbird migration she discovered, up to three thousand ruby-throats on peak autumn days, is unparalleled elsewhere.

The migration was not well known outside Rockport until recently. Jesse Grantham, an Audubon wildlife sanctuary manager, knew of Mrs. Hagar's writings. He recalled, "In the fall of 1988, while with Rockport birders Dan and Betty Baker in Bayside, we discovered several homes where residents were maintaining hummingbird feeders. There were literally hundreds of hummingbirds swarming about these artificial food sources. It was so spectacular I began to wonder how we could share this event with others," he continued. "That day, the idea of a Hummingbird Celebration was created."

The goal of the annual event is to encourage appreciation of hummingbirds by understanding what they feed upon, why they migrate, how to preserve natural areas, and how to create a backyard habitat to attract them.

Migrating ruby-throated hummingbirds are the focus of the Hummer/Bird Celebration held each autumn in Rockport, Texas. (Photo copyright by Steve and Dave Maslowski)

My Gulf Coast hummingbird odyssey began on a September afternoon as I drove from the Corpus Christi airport along the thirty-mile rural route to Rockport. I watched a lingering, rosy sunset reflected in clouds billowing over the Gulf of Mexico. When I commented on the picturesque sunset to my host at a local motel, she replied that after a recent ten-inch rainstorm, everyone was happy to see the clouds departing.

I awoke the following morning to the aroma of salt air wafting on a warm sea breeze and sounds of laughing gulls along the beach, two blocks from my motel. Rockport (population 6,300) retains the air of a mellow fishing village, even though it now caters to a growing number of birders and water-oriented vacationers.

When I arrived at the first workshop, I found 350 seats facing a stage decorated with potted plants attractive to hummingbirds. Many of the chairs were already occupied by casually dressed folks of all ages, some with binoculars still dangling around their necks. More milled around booths touting books, feeders, carvings, and nature tours, which lined three sides of the auditorium and spilled onto a seaside patio.

Before the speakers began, Betty Baker welcomed us with typical Texas hospitality. "The Hummer/Bird Celebration began in 1989 with about two hundred attendees," she proclaimed. "Last year we jumped to six hundred. Y'all know that a bird feeder is only as good as the birds it draws in. Well, this year we must have a great program because we expect about 1,500 participants.

"As you know," she continued, "the numbers of hummingbirds arriving for our festival are not as predictable. The first migrant ruby-throats show up in Rockport around the end of July. More and more arrive as weeks pass, peaking in early September and tapering off through October. The dates for the Hummer/Bird Celebration are set in hopes of capturing that window of peak viewing opportunity. The birds seem to come in waves, with one individual probably only staying a day or two."

"I hate to tell you," Betty apologized, "but with the stormy weather, most of the birds that were here have gone. We may not have the concentrations this week that you will see in video tapes of past celebrations.

"But I'm sure you'll agree," she continued, "that to witness two hundred, fifty, even one of these small birds at a feeder or a nectar-filled flower on this long migratory flight is humbling. Even though the birds' timing might not match ours, their fortitude is what it's all about."

Later, I viewed a video filmed at a local residence the previous year. The camera was focused on five feeders along the side of a house. A buzz of fifteen to twenty hummingbirds zipped around them. "That's the way it is at my house every fall," a Fulton native told me as we watched together.

On another break I stopped at a booth sponsored by the Coastal Bend Audubon Society. Here I picked up directions for a self-guided tour of hummingbird hot-spots around Rockport and Fulton. The route included country roads where wild Turk's cap blooms in abundance. The following day, I saw and heard a dozen or more hummers at each of these spots. As I stood absorbing the sights and sounds, I could imagine Connie Hagar viewing a similar scene half a century ago.

Mrs. Hagar would undoubtedly be shocked by the beachfront development that has occurred in the decades since she birded here. Large segments of the coastal live oak forest have been bulldozed for waterfront homes. Some developers have retained the attractively gnarled oaks in their landscaping plan, but the underbrush, including nectar-laden Turk's cap, has been cleared from beneath most of them.

Several stops on the Audubon tour route were at homes where residents have landscaped with hummingbirds in mind. At Glenn and Cheryl Olsen's cottage, a bed of bright red *Pentas*, orange *Lantana*, and scarlet sage in front of the screened porch enticed hummingbirds into easy viewing range. Feeders near the porch and dangling from live oak limbs provided more incentive for hummers to stop and refresh themselves. The uncleared woods behind the house harbored additional wildflowers for the wayfarers.

Farther north along the shore of Aransas Bay

Buff-bellied hummingbirds are expanding their range along the Texas Gulf Coast. (Photo copyright by Sid Rucker)

Above: Like many residents in the Rockport-Fulton area, Fred and Laverne Boden have filled their yard with flowers, shrubs, and feeders to entice hummingbirds. (Photo copyright by Connie Toops)

Left: The bright, nectar-laden blossoms of scarlet sage provide an excellent food source for hummingbirds. (Photo copyright by Connie Toops)

was the home of Curt and Anna Reemsnyder. They have planted a collage of nectar-producing flowers such as shrimp plant, Cape honeysuckle, scarlet bush, and Mexican firecracker to attract hummingbirds. Among them were beds of zinnias, *Coreopsis,* and *Gaillardia* to lure butterflies. Various sizes of hummingbird feeders hung around the yard and a mister, which the birds used for bathing, was in clear view from the patio.

Fred and Laverne Boden retired to Fulton twelve years ago. They have filled their yard with old-fashioned hibiscus shrubs, trumpet creeper and Queen's wreath vines, verbena, and shrimp plants, supplemented with numerous hummingbird feeders. In addition to hordes of ruby-throated migrants and summering black-chins, they have hosted a few overwintering rufous hummers. When they noticed a different hummingbird in the yard, they summoned Betty Baker. She identified it as a buff-bellied hummer, a Mexican species seen occasionally along the Texas coast.

I returned several times to a roadside rest stop on Highway 35, which passes through Rockport and Fulton. Testifying to awakening pride in the area's hummingbird migration, the Texas Highway Department, local service clubs, businesses, and community members have installed a demonstration garden at the rest stop. Among hummer favorites in the wide, curving flower bed are native salvia, Mexican cigar plant, Mexican honeysuckle, desert willow, and butterfly plant. Turk's cap grows beneath the live oaks and trumpet vine twines on shrubs behind the garden. Brochures are available to visitors who wish to identify the flowers or start a hummer planting in their own yards.

If the numbers of hummingbird feeders and landscaping plants purchased at the Hummer/Bird Celebration were any indication, the goal of inspiring people to create their own backyard habitat for wildlife was met. Although those of us who stayed only a few days did not see hummers in great concentrations, it was satisfying to know they still pass through the area in high densities.

Jesse Grantham, originator of the Hummer/Bird Celebration, summarized the impact of Rockport's mass migration. "From the very beginning, I thought this diminutive little bird could be the vehicle for encouraging people to have a greater appreciation for nature," he said. "The response has been overwhelming, with interests going far beyond just looking at hummingbirds. People who come are discovering that the celebration is educational, thought-provoking, and fun. These educated folks are going to be the ones to make future decisions on what happens to our country's wildlife."

Attracting Hummingbirds

Imagine yourself outside on a sunny spring day. Bright flowers and shrubs are blooming around you and water trickles across some rocks into a small pool. Suddenly you hear a familiar whirring sound. A glittering hummingbird hovers like Tinkerbell, watching you inquisitively. Then it darts off to explore the surrounding blossoms.

The setting for this scenario could be a park or botanical garden, but it could also be your own back yard. Gardening and feeding birds are two of the most popular American pastimes. It is very easy to install a hummingbird feeder outside a picture window or near your patio. With a little more effort you can create a backyard habitat designed especially to attract these fascinating and colorful elfin sprites. The moments you spend supplying the needs of hummingbirds will be repaid many times over in the delight of watching their feisty and energetic acrobatics.

One of the earliest hummingbird feeders was constructed a century ago. A Massachusetts watercolor artist cut a lifelike trumpet creeper from her painting and tied it to a small vial of sugar water. To her delight, resident ruby-throated hummingbirds visited fearlessly. Two Iowa sisters read of this feeder in *Bird-Lore* magazine and tried their own version, using red and yellow flowers snipped from oilcloth. At first, they attached them to bottles of sugar water. Eventually their ruby-throats learned to eat from undecorated bottles.

Within the past two decades, many styles of hummingbird feeders have been marketed. Choosing the best design for your needs depends in part on where you live and how many birds you will lure. Feeders come in two general types—bottles inverted to create a vacuum, with nectar available at one or more ports around the bottom, or saucer-shaped containers with holes in the top. Each style has strengths and weaknesses.

Some inverted bottle feeders slosh and drip in the wind. In direct sun, air inside expands, pushing nectar out the ports. The spilled liquid attracts bees and wasps. Saucer-shaped feeders do not drip, but certain models are hard to clean. If you live where it rains frequently, saucers catch precipitation, which dilutes the nectar. Dirt and small insects also drop into the holes. Some feeders come with a dome-shaped roof. This shades the feeder and sheds rain, but where it is windy, the roof catches breezes and sloshes the feeder.

Perches are not necessary. They do encourage hummingbirds to stay longer at each feeding, and perches allow human viewers to see the birds at rest. Since hummingbirds zip from flower to flower so rapidly, it is a treat to observe them in quiet moments. A feeder mounted near a window allows close visual access without disturbing the birds.

Perches permit other birds, such as orioles, finches, and woodpeckers, to hang on the feeder and steal nectar. If you are attracting large nectar-eaters, make or purchase a feeder for them so they will leave the hummingbird feeder alone.

Ants are persistent nectar thieves. Troops of

Planting flowers attractive to hummingbirds will bring these energetic Tinkerbells to your doorstep. (Photo copyright by Luke Wade)

them will march up a tree or along the eaves of a house to reach a feeder. They can be discouraged by placing salad oil or Tanglefoot, a sticky substance available at garden supply stores, on support wires. Commercial ant traps, small reservoirs filled with water or grease that hang above the feeder, also work well.

Bees and wasps can be discouraged with bee guards or screening that fits over the nectar ports. Guard openings must be at least 1/16-inch wide to allow hummingbirds to insert their beaks. Feeders with decorative flowers that lie flat against the base (such as Perky-Pet "No Drip" models) seem less attractive to insects. Some people smear Vasoline or salad oil on the ports to repel bees and wasps. Using a damp cloth to keep sloshed nectar cleaned off the ports also deters visits from competing insects. Hummingbirds will readily dine on gnats and fruit flies. It is easy to attract these beneficial insects by placing banana peels or melon rinds in an inconspicuous place near nectar feeders.

Ease of cleaning is a basic consideration when purchasing a commercial feeder. Each time sugar-water is replaced, the feeder should be thoroughly scrubbed. This includes disassembling the reservoir and ports. Use a bottle brush or old toothbrush to clean any mold or fungus that has accumulated. Hot water with a bit of vinegar or chlorine bleach added is an effective cleanser. All parts of the feeder should be rinsed thoroughly before reassembly. Hanging bottles are easy to take apart. Their glass reservoirs are sturdy and can be sterilized in a dishwasher. Plastic parts on some less expensive feeders fade and crack in prolonged exposure to the weather.

Nectar will ferment when left in a feeder too long. A hummingbird sipping spoiled nectar is like a human taking a big gulp of sour milk—neither wants to drink more! In cool climates, nectar should be replaced every three to five days. Above 70 degrees Fahrenheit, or for feeders in direct sunlight, nectar should be changed every two days. If the solution appears cloudy, has a sour odor, or if

Prune carefully during the spring and summer. Hummingbirds may nest among your landscape plantings. (Photo copyright by Clayton A. Fogle)

hummers seem to avoid the feeder, the sugar water is probably spoiled. When you have a large feeder and only a few hummers drinking from it, the nectar may ferment before it is all consumed. Consider using a smaller model or only partly filling the large one.

Syrup sweeter than one part sugar to four parts water should not be offered. Although hummers will eat more concentrated mixtures, higher sugar content makes them thirsty. If they have no other source of water, they will drink more solution and get thirstier. In zoo-raised birds, prolonged ingestion of highly concentrated sugar solutions is thought to cause liver damage.

Commercial nectar mixes are available, but it is easy to make your own hummingbird feeding solution. Use one cup sugar dissolved in four cups of water. Heating the water speeds the mixing process and helps prevent fermentation. Some people prefer to place sugar and water in a kettle on the stove and bring it to a boil. Be aware that in the process, water will evaporate from an open pan, making the sugar solution more concentrated. You may wish to add a little extra water if using this method.

I measure sugar into a large glass container, boil water, add it to the sugar, and stir rapidly until the sugar dissolves. I often make a large batch and refrigerate the excess syrup, which keeps well for at least a week. Freshly mixed syrup should be cooled before offering it to hummers.

Honey is not a good food because it spoils quickly and hosts a fungus that fatally infects hummingbird tongues, causing death by starvation. Saccharine and sugar substitutes contain no nutrition and are harmful to hummingbirds.

Red dye is included in some commercial nectars. This is controversial because the Food and Drug Administration recalled Red Dye #2 in 1975 when it was discovered unsafe for humans. The red dye now in use is certified safe for people but has not been authoritatively tested on hummingbirds. Some people like the ease of monitoring red liquid in feeders. Others feel dye is unnecessary since most feeders have red on them. If not, color can be added by painting, using tape, or flagging to make the feeder more conspicuous.

I have offered commercial red nectar, "lite" nectar (which is pink), and homemade solution side-by-side in feeders. The birds showed a slight preference for plain sugar water in my unscientific tests, but also consumed both commercial nectars.

When unlimited supplies of nectar are available at feeders, humans may think it silly for hummingbirds to waste their energy fighting over it. But defensive behavior is an instinctive carryover from protecting wildflowers. Aggressive behavior tends to be more pronounced early in the season. One way to circumvent this behavior is to maintain several feeders. If possible, put each feeder out of sight from the others. Otherwise, place a cluster of three or more feeders at each location.

If hummers are using just the feeder with no other floral nectar, a quart of 1:4 solution contains enough sugar to sustain 120 hummingbirds the size of a ruby-throat for one day. If no solution is lost to insects, bats, or spillage, a quart feeder refilled every other day sustains sixty hummers. A quart feeder filled every third day supports about forty birds.

At least one species of hummingbird is found throughout most of the United States. On the West Coast or in the Southwest, you can expect several varieties. You will be most likely to lure hummers if you have a feeder waiting for them during spring migration. Typical arrival dates for ruby-throats are: Gulf Coast/early to mid-March, mid-Atlantic states/mid-April, Midwestern states/late April, and New England or Canada/early May. Black-chin arrival dates are: Texas/early March, Colorado/early May, and Montana/mid-May. Local arrivals for migrant rufous hummers include: southern California/early February, Oregon and Washington/early March, British Columbia/early April, and Alaska/mid-April. For a short period in the spring, the combination of migrants and residents may swell local hummer populations to twice their normal numbers.

Southbound migrants also benefit from feeders left out for them. The last ruby-throats depart Canada by mid-September, pass through the Midwest in late September, and leave the Gulf Coast by mid-October. Black-chins head south by late August, passing through Texas in late October.

Above: In order for hummingbirds to bathe in pools, the water should be shallow enough around the edges for them to stand up. (Photo copyright by Hugh P. Smith, Jr.)

Right: Choosing feeders that do not slosh nectar on windy days helps reduce unwanted insects. (Photo copyright by Connie Toops)

Left: Hummingbirds readily learn to drink from feeders. The proper ratio for homemade nectar is one cup of sugar dissolved in four cups of water. (Photo copyright by Hugh P. Smith, Jr.)

Above: Hummingbirds seem to enjoy dripping water. They bathe during and after rainshowers by splashing against wet leaves. (Photo copyright by Hugh P. Smith, Jr.)

Hummingbirds often dart into the spray of garden hoses and lawn sprinklers and then perch on a nearby branch to preen. (Photo copyright by Connie Toops)

Rufous leave the northern mountains in late July or August, reaching the latitude of New Mexico by early September and departing the Southwest by November. A good rule of thumb is to maintain your feeder at least a week after you notice the last hummingbird of autumn.

In addition to providing feeders, one of the best ways to entice hummingbirds to your yard is to create habitat that is attractive to them. This includes trees or shrubs that offer perches in the sun and shade, a variety of blooming flowers, and a source of water.

Hummingbirds enjoy dripping water, which they choose over still pools. It is possible to install a saddle valve on an outdoor faucet to direct a trickle of water through plastic tubing, ultimately spraying it into the air or dribbling into a pool. If you offer a pool, make sure the edges are shallow enough for a hummer to stand in without being submerged.

At our northeastern California home, we see hummingbirds darting into the spray of sprinklers when we water the lawn. They hover momentarily, sometimes returning twice or three times for a thorough dousing. They also bathe after rain, rubbing and splashing against wet tree leaves. Then they fly to a nearby tree to preen. The grooming ritual includes a thorough scratching of head and beak areas with the foot, which is raised over the top of the wing. Hummers pull primary feathers lightly through their beaks to realign the barbules.

In spring you can offer female hummingbirds raw materials for their nests. They are fond of strands of human hair and lint from clothes driers, especially in shades of gray, white, or blue-gray. These materials can be placed in an empty suet container (made of large-gauge hardware cloth) on a tree trunk.

There are two possibilities to consider if your properly maintained feeder is not attracting hummingbirds. Habitat near your yard may be so lush with wild sources of nectar that hummingbirds do

Left: Shrubs and perennials with numerous red flowers that bloom over a lengthy period are excellent choices for back-yard hummingbird landscapes. (Photo copyright by Connie Toops)

Above: Time invested in landscaping or maintaining feeders for hummingbirds will be amply repaid in the delight of watching the birds' acrobatic flights. (Photo copyright by Connie Toops)

not need extra food. Hummers will desert feeders during periods when honeysuckle, mimosa, trumpet creeper, citrus trees, or rich local wildflowers are especially abundant. It is more likely, however, that areas surrounding your backyard are barren of flowers, water, and shelter necessary to attract hummers.

To improve your wildlife landscape, begin by planting a tree or two as a centerpiece. Add a few shrubs, perennials, and annual flowers. Annuals give immediate sources of nectar but live only one season. Perennials require less prolonged maintenance and bloom every year. Shrubs and trees take longer to mature but provide shade, perches, and in some cases, flowers.

If space is limited, consider planting floral vines, such as honeysuckles, that will climb on fences, trellises, or walls. Even a window box or hanging basket can attract hummers when planted with favorites such as impatiens, fuchsia, or salvia. When choosing plants, check with a nursery or garden center for varieties that do well in your climate and soil type. Locate tall flowers at the rear of beds and shorter ones in front, giving hummers room to maneuver. Try to have at least one or two species in bloom throughout the season.

Plants with numerous small flowers blooming over a long period are better for hummers than a few large flowers in bloom a short time. Red color, tubular shape, and abundant nectar are desirable characteristics. Native plants tend to have greater tolerance to local weather conditions. They usually produce more sugar than hybrid varieties. Many hybrid roses, geraniums, and petunias are showy but offer little nectar. Individual birds investigate them but will not return regularly to feed.

Study the plants your hummingbirds prefer. Replace less productive species with more enticing ones. Ask at your garden center about hardy varieties requiring little maintenance or specialized care. Fertilize periodically in the growing season, but avoid using insecticides, fungicides, or other chemicals that could harm hummers. Watch for nests if you prune shrubs and trees during the late spring or early summer.

The following tables include examples of native and landscaping plants attractive to hummingbirds in various regions of the United States and southern Canada. Contact a nursery, garden club members, or the National Wildlife Federation with specific questions about developing a landscape plan for your backyard wildlife.

You may not attract a hummingbird the first day, week, or in some cases, even the first month you put up a feeder. But once established, attractive habitats will bring hungry hummingbirds to your doorstep year after year.

Plants for Regional Hummingbird Gardens

CENTRAL/SOUTHERN CALIFORNIA
Hummingbirds: Allen's, Anna's, black-chinned, calliope, Costa's, rufous

ANNUALS
impatiens	(Impatiens hybr.)
toadflax	(Linaria maroccana)
scarlet sage	(Salvia splendens)

PERENNIALS
aloe vera	(Aloe perryi)
crimson columbine	(Aquilegia formosa)
coast paintbrush	(Castilleja affinis)
Indian paintbrush	(Castilleja californica)
scarlet larkspur	(Delphinium cardinale)
fuchsia 'Marinka,' 'red spider,' 'June bride,' 'gartenmeister'	(Fuchsia hybr.)
shrimp plant	(Justica brandgeana)
scarlet monkeyflower	(Mimulus cardinalis)
scarlet bugler	(Penstemon centranthifolius)
scarlet penstemon	(Penstemon labrosus)
pineapple sage 'ruby'	(Salvia elegans)
California Indian pink	(Silene californica)
California fuchsia	(Zauschneria californica)

VINES
trailing abutilon	(Abutilon megapotamicum)
trumpet creeper	(Campsis radicans)
blood-red trumpet vine	(Disticus buccinatoria)
trumpet honeysuckle	(Lonicera sempervirens)

SHRUBS
Chinese lantern	(Abutilon hybridum)
manzanita 'Dr. Hurd'	(Arctostaphylos hybr.)
red bird-of-paradise	(Caesalpinia pulcherrima)
pink powder puff	(Calliandra haematocephala)
crimson bottlebrush	(Callistemon citrinus)
ocotillo	(Fouquieria splendens)
chuparosa	(Justica californica)
lantana	(Lantana camara)
lion's ear	(Leonotis leonurus)
fuchsia-flowering gooseberry	(Ribes speciosum)
Cape honeysuckle	(Tecomaria capensis)

TREES
mimosa	(Albizzia julibrissin)
orange, lemon, grapefruit	(Citrus sp.)
red-flowering eucalyptus	(Eucalyptus sp.)
silk oak	(Grevillea robusta)
tree tobacco	(Nicotiana glauca)

Flowering maple, trailing abutilon, and Chinese lantern—all types of hibiscus—lure hummingbirds into California gardens. (Photo copyright by Connie Toops) Inset: Saucer-shaped feeders work well in climates that are dry and windy. Before purchasing this style, make certain the feeder comes apart for easy cleaning. (Photo copyright by Clayton A. Fogle)

PACIFIC NORTHWEST

Hummingbirds: Anna's, black-chinned, calliope, rufous

ANNUALS

satin flower	(*Clarkia amoena*)
grand collomia	(*Collomia grandiflora*)
foxglove	(*Digitalis purpurea*)
gladiolus	(*Gladiolus* hybr.)
velvetflower	(*Salpiglossis sinuata*)
nasturtium	(*Tropaeolum majus*)

PERENNIALS

crimson columbine	(*Aquilegia formosa*)
paintbrush	(*Castilleja* sp.)
delphinium	(*Delphinium hybridum*)
bleeding heart	(*Dicentra formosa*)
fireweed	(*Epilobium angustifolium*)
red-hot poker	(*Kniphofia uvaria*)
Columbia lily	(*Lilium columbianum*)
Lewis' monkeyflower	(*Mimulus lewisii*)
bird's beak lousewort	(*Pedicularis ornithorhyncha*)
penstemon	(*Penstemon* sp.)
hedge-nettle	(*Stachys cooleyae*)

VINES

trumpet creeper	(*Campsis radicans*)
Chilean bellflower	(*Laphgeria rosea*)
orange honeysuckle	(*Lonicera ciliosa*)
scarlet runner bean	(*Phaseolus coccinea*)

SHRUBS

manzanita	(*Arctostaphylos* sp.)
hardy fuchsia 'riccartonii'	(*Fuchsia magellanica*)
twinberry	(*Lonicera invulcrata*)
rhododendron	(*Rhododendron macrophyllum*)
red-flowering currant	(*Ribes sanguineum*)
fuchsia-flowering gooseberry	(*Ribes speciosum*)
red elderberry	(*Sambucus racemosa*)

TREES

California buckeye	(*Aesculus californica*)
horsechestnut	(*Aesculus hippocastanum*)
Pacific madrone	(*Arbutus menziesii*)

Nectar-rich fireweed graces meadows and stream banks of the Pacific Northwest. (Photo copyright by Connie Toops)

Above inset: Fuchsias, which are native to the cool mountains of Central and South America, are readily visited by hummingbirds. (Photo copyright by Connie Toops)

Below inset: Rufous hummers range the farthest north of all hummingbirds. They reach southeastern Alaska and the southern Yukon each summer. (Photo copyright by Clayton A. Fogle)

NORTHEAST
Hummingbird: ruby-throated

ANNUALS
snapdragon	*(Antirrhinum majus)*
foxglove	*(Digitalis purpurea)*
gladiolus	*(Gladiolus* hybr.*)*
impatiens	*(Impatiens* hybr.*)*
scarlet sage	*(Salvia splendens)*
nasturtium	*(Tropaeolum majus)*

PERENNIALS
columbine	*(Aquilegia canadensis)*
butterflyweed	*(Asclepias tuberosa)*
delphinium 'King Arthur'	*(Delphinium hybridum)*
bleeding heart	*(Dicentra eximia)*
coral bells	*(Heuchera sanguinea)*
jewelweed	*(Impatiens capensis)*
cardinal flower	*(Lobelia cardinalis)*
bee-balm	*(Monarda didyma)*
wild sweet William	*(Phlox maculata)*

VINES
morning glory	*(Ipomoea purpurea)*
trumpet honeysuckle	*(Lonicera sempervirens)*
scarlet runner bean	*(Phaseolus coccinea)*

SHRUBS
buttonbush	*(Cephalanthus occidentalis)*
flowering quince	*(Chaenomeles japonica)*
autumn olive	*(Elaeagnus umbellata)*
Tatarian honeysuckle	*(Lonicera tatarica)*
lilac	*(Syringa vulgaris)*
cardinal shrub	*(Weigela florida)*

TREES
redbud	*(Cercis canadensis)*
hawthorn	*(Crategus* sp.*)*
tulip tree	*(Liriodendron tulipifera)*

Impatiens are colorful annual flowers that attract humming-birds. They grow well in the sun or partial shade. (Photo copyright by Connie Toops)

Left inset: Jewelweed, also called touch-me-not because of its popping seed pods, is a common wildflower of wet, shady areas throughout the East and Midwest. (Photo copyright by Connie Toops)

Right inset: Male ruby-throated hummingbird. (Photo copyright by Luke Wade)

SOUTHEAST
Hummingbird: ruby-throated

ANNUALS
sweet William	*(Dianthus barbatus)*
foxglove	*(Digitalis purpurea)*
flowering tobacco	*(Nicotiana alata)*
scarlet sage	*(Salvia splendens)*

PERENNIALS
Indian paintbrush	*(Castilleja coccinea)*
red turtlehead	*(Chelone obliqua)*
bleeding heart	*(Dicentra eximia)*
jewelweed	*(Impatiens capensis)*
standing cypress	*(Ipomopsis rubra)*
cardinal flower	*(Lobelia cardinalis)*
bee-balm	*(Monarda didyma)*
fire pink	*(Silene virginica)*
Indian pink	*(Spigelia marilandica)*

VINES
cross vine	*(Bignonia capreolata)*
trumpet creeper	*(Campsis radicans)*
red morning glory	*(Ipomoea coccinea)*
cypress vine	*(Ipomoea quamoclit)*
trumpet honeysuckle	*(Lonicera sempervirens)*

SHRUBS
flowering quince	*(Chaenomeles japonica)*
rose of Sharon	*(Hibiscus syriacus)*
lantana	*(Lantana camara)*
azalea	*(Rhododendron* sp.*)*
coralberry	*(Symphoricarpos orbiculatus)*

TREES
red buckeye	*(Aesculus pavia)*
catalpa	*(Catalpa bignonioides)*
tulip tree	*(Liriodendron tulipifera)*
crab apple	*(Malus coronaria)*
black locust	*(Robinia pseudoacacia)*

Salvia is available at most nurseries. Its vibrant red flowers are hummingbird favorites. (Photo copyright by Connie Toops)

Above inset: Large flowers of the trumpet creeper vine offer ten times more nectar than most other blossoms hummingbirds visit. (Photo copyright by Connie Toops)

Below inset: In addition to food and water, backyard landscapes should offer perches and sheltered roosts for hummingbirds. (Photo copyright by Tom Pawlesh)

FLORIDA/GULF COAST
SOUTHERN TEXAS
Hummingbirds: black–chinned (TX), buff–bellied (TX), ruby–throated, rufous (TX)

ANNUALS

garden balsam	*(Impatiens balsamina)*
impatiens	*(Impatiens hybr.)*
lemon mint	*(Monarda citriodora)*
annual phlox	*(Phlox drummondii)*
scarlet sage	*(Salvia coccinea, S. splendens)*

PERENNIALS

Mexican butterflyweed	*(Asclepias curassivica)*
canna	*(Canna inidca)*
Mexican cigar plant	*(Cuphea micropetala)*
bromeliads	*(Tillandsia, Guzmania)*
scarlet hibiscus	*(Hibiscus coccineus)*
standing cypress	*(Ipomopsis rubra)*
shrimp plant	*(Justica brandegeana)*
firecracker bush	*(Russelia equisetiformis)*
autumn sage	*(Salvia greggii)*
Mexican bush sage	*(Salvia leucantha)*

VINES

trumpet creeper	*(Campsis radicans)*
Carolina jasmine	*(Gelsemium sempervirens)*
red morning glory	*(Ipomoea coccinea)*
trumpet honeysuckle	*(Lonicera sempervirens)*

SHRUBS

flame flower	*(Anisacanthus wrightii)*
purple cestrum	*(Cestrum elegans)*
coral bean	*(Erythrina herbacea)*
scarlet bush	*(Hamela patens)*
Mexican honeysuckle	*(Justica spicigera)*
calico bush	*(Lantana horrida)*
Mexican Turk's cap	*(Malvaviscus arboreus)*
Turk's cap	*(Malvaviscus drummondii)*
red pentas	*(Pentas lanceolata)*

TREES

desert willow	*(Chilopsis linearis)*
orange, lemon, grapefruit	*(Citrus sp.)*
royal poinciana	*(Delonix regia)*
loquat	*(Eriobotrya japonica)*

Subtropical bromeliads, such as this Guzmania, attach to tree bark and branches. They draw moisture and nutrition through aerial roots. (Photo copyright by Connie Toops)

Left inset: The bright flowers of coral bean are borne on small shrubs. (Photo copyright by Connie Toops)

Right inset: Black-chinned hummingbirds breed in central and western Texas. They are occasionally seen along the northern Gulf Coast and in Florida, especially during autumn migration. (Photo copyright by Luke Wade)

WESTERN MOUNTAINS
Hummingbirds: black-chinned, broad-tailed, calliope, rufous

ANNUALS
snapdragon	*(Antirrhinum majus)*
tiny trumpet	*(Collomia linearis)*
impatiens	*(Impatiens hybr.)*
scarlet sage	*(Salvia splendens)*
nasturtium	*(Tropaeolum majus)*

PERENNIALS
columbine	*(Aquilegia elegantula)*
Wyoming paintbrush	*(Castilleja linariaefolia)*
giant red paintbrush	*(Castilleja minata)*
fireweed	*(Epilobium angustifolium)*
scarlet gilia	*(Ipomopsis aggregata)*
scarlet lychnis	*(Lychnis chalcedonica)*
horsemint	*(Monarda fistulosa)*
penstemon	*(Penstemon bridgesii)*
longleaf phlox	*(Phlox longifolia)*

VINES
morning glory	*(Ipomoea purpurea)*
orange honeysuckle	*(Lonicera ciliosa)*
trumpet honeysuckle	*(Lonicera sempervirens)*

SHRUBS
green–leaf manzanita	*(Arctostaphylos patula)*
Siberian pea	*(Caragana arborescens)*
flowering quince	*(Chaenomeles japonica)*
twinberry	*(Lonicera involucrata)*
flowering currant	*(Ribes gordonianum)*
blueberry	*(Vaccinium sp.)*

TREES
catalpa	*(Catalpa speciosa)*
Russian olive	*(Elaeagnus angustifolia)*
flowering crab	*(Malus sp.)*
chokecherry	*(Prunus virginiana)*
black locust	*(Robinia pseudoacacia)*

Flower-filled mountain meadows in the West are excellent places to observe broad-tailed, calliope, and rufous humming-birds. (Photo copyright by Connie Toops)

Left inset: A broad-tailed hummingbird sips nectar from horsemint. (Photo copyright by Charles W. Melton)

Right inset: Numerous species of penstemon grow in the western mountains. Although red-flowered varieties may attract a hummingbird's attention more rapidly, the tiny birds readily learn to visit nectar-laden blue and purple flowers. (Photo copyright by Connie Toops)

SOUTHWEST

Hummingbirds: Allen's, Anna's, black-chinned, blue-throated, broad-billed, broad-tailed, calliope, Costa's, Lucifer, magnificent, rufous, violet-crowned, white-eared

ANNUALS
scarlet flax	*(Linum grandiflorum)*
scarlet betony	*(Stachys coccinea)*

PERENNIALS
bouvardia	*(Bouvardia ternifolia)*
cardinal flower	*(Lobelia cardinalis)*
scarlet monkeyflower	*(Mimulus cardinalis)*
beardlip penstemon	*(Penstemon barbatus)*
scarlet bugler	*(Penstemon centranthifolius)*
Eaton's penstemon	*(Penstemon eatonii)*
Parry's penstemon	*(Penstemon parryi)*
Arizona penstemon	*(Penstemon pseudospectabilis)*
autumn sage	*(Salvia greggii)*
Lemmon's sage	*(Salvia lemmonii)*
mountain sage	*(Salvia regla)*
Mexican pink	*(Silene laciniata)*

VINES
snapdragon vine	*(Asarina antirrhiniflora)*
trumpet creeper	*(Campsis radicans)*
glory flower	*(Eccremocarpus scaber)*
red morning glory	*(Ipomoea cristulata)*
cypress vine	*(Ipomoea quamoclit)*

SHRUBS
desert honeysuckle	*(Anisacanthus thurberi)*
red bird-of-paradise	*(Caesalpinia pulcherrima)*
bladderbush	*(Cleome isomeris)*
hedgehog cactus	*(Echinocereus triglochidiatus)*
ocotillo	*(Fouquieria splendens)*
chuparosa	*(Justica californica)*
yellow trumpetflower	*(Tecoma stans)*

TREES
coral bean	*(Erythrina flabelliformis)*
eucalyptus	*(Eucalyptus sp.)*

Summer monsoon rains bring a "second spring" burst of wildflowers to the mountains of the Southwest. Late-nesting residents and southbound migrant hummingbirds abound here in August. (Photo copyright by Connie Toops)

Left inset: Cypress vine, a member of the morning glory family, is easy to grow on a backyard trellis. (Photo copyright by Connie Toops)

Right inset: Among the most spectacular of the southwestern hummers is the aptly named magnificent hummingbird. (Photo copyright by Connie Toops)

MIDWEST

Hummingbird: ruby-throated

ANNUALS

hollyhock	*(Alcea rosea)*
spider flower	*(Cleome spinosa)*
sweet William	*(Dianthus barbatus)*
foxglove	*(Digitalis purpurea)*
gladiolus	*(Gladiolus hybr.)*
flowering tobacco	*(Nicotiana alata)*
scarlet sage	*(Salvia splendens)*

PERENNIALS

columbine	*(Aquilegia canadensis)*
butterflyweed	*(Asclepias tuberosa)*
Indian paintbrush	*(Castilleja coccinea)*
bleeding heart	*(Dicentra spectabilis)*
coral bells	*(Heuchera sanguinea)*
jewelweed	*(Impatiens capensis)*
cardinal flower	*(Lobelia cardinalis)*
bee-balm	*(Monarda didyma)*
fire pink	*(Silene virginica)*

VINES

trumpet creeper	*(Campsis radicans)*
morning glory	*(Ipomoea purpurea)*
trumpet honeysuckle	*(Lonicera sempervirens)*

SHRUBS

butterfly bush	*(Buddleia davidii)*
buttonbush	*(Cephalanthus occidentalis)*
flowering quince	*(Chaenomeles japonica)*
rose of Sharon	*(Hibiscus syriacus)*
Tatarian honeysuckle	*(Lonicera tatarica)*
coralberry	*(Symphoricarpos orbiculatus)*
lilac	*(Syringa vulgaris)*

TREES

Ohio buckeye	*(Aesculus glabra)*
horsechestnut	*(Aesculus hippocastanum)*
tulip tree	*(Liriodendron tulipifera)*
crab apple	*(Malus angustifolia)*
black locust	*(Robinia pseudoacacia)*

Index

Attracting hummingbirds, 5, 7,
    31, 68, 69, 85, 90, 91, 93,
    96, 99, 101–124
Banding, 37, 71–77
Bathing, 107
Beak, see Bill
Bill, 27, 31, 33, 35, 38–40, 55, 73
Breeding, 43, 45, 49, 51
Chicks, 9, 53, 55
    described, 55, 56
Color, 9, 16, 24–25, 89
Courtship, 25, 43–49
    see also Leks
Crop, 27
Development, 55–56, 59
Diet, see Food
Digestion, 27, 30–31
Distribution and range, 9, 10,
    13, 68–69, 93, 104, 107
Eggs, 51, 53, 80
    described, 9, 53, 55
    hatching, 80
    incubation, 55
Endangered species, 17
Energy requirements, 9, 20, 30–
    31, 33, 41
Evolution, 9, 10, 11
Feathers, 19–25, 46
Feeding,
    flights, 31
    frenzies, 7
    young, 55, 56
    see also Food

Feeders, 67–68, 85, 96, 99, 101,
    104, 107, 109
Flight, 9, 19–20, 45
    speed of, 20, 22
Flowers attractive to humming-
    birds, 31, 33, 109–124
    adaptations of, 31, 33, 38–39
    see also Attracting humming-
        birds
Food, 27–41, 67–69, 104
    at sapsucker drillings, 35
    competition with insects for,
        31, 33, 40, 89, 103
    competition with other hum-
        mingbirds for, 35
    daily intake, 31
    (insects), 10, 30–31, 56
    (nectar), 27, 31, 33, 35, 38, 39,
        40, 43, 56, 103–104
Gorget, 25, 45, 79, 93
Habitat, 10, 13, 17, 38, 84, 91,
    93
    loss of, 17, 96
Heart rate, 9, 20
    during torpor, 41
Hovering, 20, 33
Hummingbird mites, 33, 35
Hybrids, 45, 80
Landscaping for hummingbirds,
    see Attracting humming-
    birds; Flowers attractive to
    hummingbirds

Legends, 13, 16, 41
Leks, 46, 49
Life spans, 59
    see also Mortality
Metabolism, 9, 27, 40
Migration, 10–11, 37, 43, 61–69,
    77, 79, 89, 93, 95–96, 104,
    107
Mortality, 56, 59, 64
    see also Life spans
Nesting, 10, 59
Nests, 9, 45, 51–59, 80, 84, 109
    building, 43, 45, 51, 59
    locations of, 51, 53
    materials for, 51, 53, 109
    size, 51, 53
Plumage, 16, 22, 45, 49
    iridescence, 24, 25, 79, 93
    see also Feathers
Predators, 25, 59
Size, 9
Sleep, 5
Tail, 25, 43, 45–46
Territories, 37, 43, 45, 89
    defense of, 43, 45, 104
Tongue, 27
Torpor, 40–41, 55
Voice, 7, 45, 46, 49, 75, 80, 89,
    93
Weight, 9, 22, 67–68, 73
Wings, 19–20, 25

Ruby-throated hummingbirds frequently feed on columbine nectar. (Photo copyright by Steve and Dave Maslowski) Inset: Cardinal flower, which grows in damp areas, blooms in late summer. (Photo copyright by Pat Toops)

References

Berger, Bruce. *A Dazzle of Hummingbirds*. San Luis Obispo, CA: Blake Publishing, 1989.

Brice, A.T., K.H. Dahl, and C.R. Grau. "Pollen Digestibility by Hummingbirds and Psittacines." *Condor* 91 (1989): 681–688.

Brandt, Herbert. *Arizona and Its Birdlife*. Cleveland: Bird Research Foundation, 1951.

Calder, William A. III. "Southbound through Colorado: Migration of Rufous Hummingbirds." *National Geographic Research* 3(1) (1987): 40–51.

Calder, William A. III, and Elinor G. Jones. "Implications of Recapture Data for Migration of the Rufous Hummingbird in the Rocky Mountains." *Auk* 106 (July 1989): 488–489.

Coastal Bend Audubon Society. "Hummingbird Plants for South Texas." Corpus Christi, TX: (pamphlet), 1991.

Colwell, Robert K. "Stowaways on the Hummingbird Express." *Natural History* 94 (July 1985): 56–63.

Dennis, John V., and Pat Murphy. "Special Reprint I: Hummingbirds." *Bird Watcher's Digest* 1988.

Feinsinger, I.W. "The Long and Short of Hummingbird Bills." *International Wildlife* 18 (July 1988): 14–17.

Ferry, David R., and Linda H. Ferry. "Southeastern Arizona Hummingbird Project." *Zoonooz* 62 (July 1989): 12–16.

Goldsmith, Kenneth M., and Timothy H. Goldsmith. "Sense of Smell in the Black-chinned Hummingbird." *Condor* 84 (1982): 237–238.

Goldsmith, Timothy H., and Kenneth M. Goldsmith. "Discrimination of Colors by the Black-chinned Hummingbird." *Journal of Comparative Physiology* 130 (1979): 209–220.

Greenewalt, Crawford H. *Hummingbirds*. New York: Dover, 1990.

Gretch, Mark. "The Early Hummingbird." *Bird Watcher's Digest* 11 (May 1989): 84–86.

Harrison, George H. "How to Attract Hummingbirds." *National Wildlife* 23 (April 1985): 42–44.

Heibert, Sara M. "Energy Costs and Temporal Organization of Torpor in the Rufous Hummingbird." *Physiological Zoology* 63(6) (1990): 1082–1097.

Holmgren, Virginia C. *The Way of the Hummingbird*. Santa Barbara, CA: Capra Press, 1986.

Johnsgard, Paul A. *The Hummingbirds of North America*. Washington, DC: Smithsonian Press, 1983.

Long, Michael E. "Secrets of Animal Navigation." *National Geographic* 179 (June 1991): 70–99.

Newfield, Nancy L. "Plant a Hummingbird Garden." *Birder's World* 1 (March 1987): 46–50.

———. "When to Take Down Your Hummingbird Feeder." *Bird Watcher's Digest* 12 (September 1989): 80–83.

———. "A Strategy for Hummers." *Birder's World* 4 (June 1990): 54–56.

———. "Hummers: Losing Ground." *Bird Watcher's Digest* 13 (July 1991): 62–68.

Newsom-Brighton, Maryanne. "A Garden Fit for Hummingbirds." *Bird Watcher's Digest* 10 (January 1988): 18–25.

Peterson, Roger Tory, and Edward Chalif. *A Field Guide to Mexican Birds*. Boston: Houghton Mifflin, 1973.

Pollock, Robert. "Summer Hummer." *Birder's World* 5 (October 1991): 30–32.

———. "The Woodland Sweetshop." *Birder's World* 5 (December 1991): 42–43.

Pyke, Graham H., and Nickolas M. Waser. "The Production of Dilute Nectars by Hummingbird and Honeyeater Flowers." *Biotropica* 13(4) (1981): 260–270.

Robbins, Chandler S., Bertel Bruun, and Herbert S. Zim. *Birds of North America*. New York: Golden, 1983.

Schultz, Lydia. "The Jewels of Summer." *Mississippi Outdoors* (July 1987): 13–16.

Scott, Shirley L., ed. *Field Guide to the Birds of North America*. Washington, DC: National Geographic, 1983.

Skutch, Alexander F. *The Life of the Hummingbird*. New York: Crown, 1973.

Spofford, Sally H. "Some Sticky Solutions." *Living Bird Quarterly* 4 (Spring 1975): 4–9.

Stallcup, Rich. "Some Thoughts on Anna's Hummingbirds and the Ten-Day Cold Snap in Middle California." *Winging It* (February 1991): 7.

Stiles, F. Gary, and Alexander F. Skutch. *A Guide to the Birds of Costa Rica*. Ithica, NY: Cornell Press, 1989.

Stokes, Donald and Lillian Stokes. *The Hummingbird Book*. Boston: Little, Brown, 1989.

————. "The Beat of a Different Hummer." *Bird Watcher's Digest* 11 (March 1989): 102–108.

————. "Hawks to Hummers." *Bird Watcher's Digest* 12 (September 1989): 108–114.

Swengel, Ann. "How to Find Hummingbirds in the Wild." *Bird Watcher's Digest* 13 (May 1991): 68–75.

Tekulsky, Mathew. *The Hummingbird Garden*. New York: Crown, 1990.

Terres, John K. *The Audubon Society Encyclopedia of North American Birds*. New York: Knopf, 1980.

Tyrell, Esther Q., and Robert A. Tyrell. *Hummingbirds: Their Life and Behavior*. New York: Crown, 1985.

————. *Hummingbirds of the Caribbean*. New York: Crown, 1990.

Vosburgh, Pat. "The Fabulous Feather." *Bird Watcher's Digest* 11 (September 1988): 60–64.

Wetmore, Alexander. *Song and Garden Birds of North America*. Washington, DC: National Geographic, 1964.

Williamson, Sheri. "Hummingbird Migration." *Wildbird* 5 (May 1991): 44–47.

Willimont, Lori A., Stanley E. Senner, and Laurie Goodrich. "Fall Migration of Ruby-throated Hummingbirds in the Northeastern United States." *Wilson Bulletin* 100(3) (1988): 482–488.

Womack, Ellie, ed. *Hummingbird Hotline*. Grove, OK: (newsletter), 1990–1992.

Overleaf: Photo of female rufous hummingbird copyright by Hugh P. Smith, Jr.